Hidden In Plain Sight 5

Andrew Thomas studied physics in the James Clerk Maxwell Building in Edinburgh University, and received his doctorate from Swansea University in 1992.

His *Hidden In Plain Sight* series of books are science bestsellers.

Also by Andrew Thomas:

Hidden In Plain Sight
The simple link between relativity
and quantum mechanics

Hidden In Plain Sight 2
The equation of the universe

Hidden In Plain Sight 3
The secret of time

Hidden In Plain Sight 4
The uncertain universe

HIDDEN PLAINSIGHT 5

Atom

ANDREW THOMAS

**AGGRIEVED
CHIPMUNK
PUBLICATIONS**

Hidden In Plain Sight 5

Copyright © 2016 Andrew D.H. Thomas

ISBN-13: 978-1519298874
ISBN-10: 1519298870

CONTENTS

PREFACE

This book is an introduction to the principles of particle physics, the physics which deals with atoms and the smallest elements of the universe. The book will emphasise the underlying principles, rather than just listing the particles (and their properties) in a rather encyclopaedic manner. I believe it is much more valuable to gain an understanding of the principles involved.

For example, we will be discovering the principle which explains why material objects feel solid to the touch, but you can't touch a ray of light. We will be discovering the real reason why positive and negative electric charge attract each other (clue: it has to do with symmetry).

I must admit, this book is perhaps more challenging than my previous books. I do think many popular science books underestimate the intelligence of their readers.

I am firmly of the conviction that if a reader is having difficulty understanding a concept then it is not because a concept is hard – it is because it is being badly explained. In this book, care is taken to describe each step simply and clearly.

Once again, thank you for your support.

Andrew Thomas (hiddeninplainsightbook@gmail.com)
Swansea, UK,
2016

"The road to a knowledge of the stars leads through the atom."

- SIR ARTHUR EDDINGTON, 1928

1

THE INVESTIGATOR

At the start of the 19th century, the land of Australia was still an unexplored country. The coastal areas had been mapped by various earlier Europeans (predominantly the Dutch, who named the country "New Holland"). However, due to the size of the continent, it was not certain if these coastal areas were connected as part of a single island, or if Australia was composed of many islands.

In July 1801, an expedition to Australia from Britain was due to set sail on the ship the *Investigator*. The main aim of the mission was to determine if Australia was a connected land mass, but, as a secondary aim, the expedition also aimed to study the flora and fauna of the island.

The expedition chose the Scottish botanist Robert Brown to be its resident naturalist. Brown, the son of a strict church minister, had learnt botany while studying in Edinburgh University where he had taken frequent trips to the Highlands of Scotland to collect and classify rare plants.

H.M. Sloop INVESTIGATOR. 1802

The *Investigator* set sail on the 18th July. It arrived in King George Sound in Western Australia six months later. As part of its voyage, the ship had to traverse the treacherous Great Barrier Reef (then called the Labyrinth). The ship eventually performed the first ever circumnavigation of Australia, thus establishing it as a continent.

During the voyage, Brown discovered approximately 2000 species of plants, almost all of which were previously unknown to science. When he returned to London three and a half years later, his reputation as a botanist of note was secured.

Brown's reputation was established on his talent for examining his Australian plant specimens in the smallest

2

detail. In order to achieve this, Brown became skilled at microscopy. Brown realised that the study of microscopic pollen grains could be used as a method to classify plants.

It was while he was using a microscope to examine some of these pollen grains suspended in water that Brown observed something rather peculiar. The pollen grains were seen to move in a jittery, random motion. This random motion was given the name *Brownian motion* (named after Robert Brown), but its origin was to remain a mystery for almost a century.

At this point, our tale takes an unexpected turn. We rejoin our tale in 1905 – the "miracle year" of Albert Einstein.

At the time, the existence of atoms was still in doubt: there had never been a direct observation of the behaviour of atoms. The stage was set for Einstein who, in one of his remarkable acts of intuition, speculated that the motion of the pollen grains might be due to the random motion of water molecules. He realised that a pollen grain was so small that – purely by chance – there would occasionally be significantly more water molecules buffeting one side of the pollen than the other. The resultant motion has been compared to a giant inflatable balloon being bounced around randomly by a crowd in a football stadium.

Einstein crushed the mathematics and realised that this effect would result in precisely the random dance of the pollen which was first observed by Robert Brown almost a century earlier.

This atomic explanation of Brownian motion represented the first directly observable effect of the kinetic theory of atoms, and is regarded as the first conclusive evidence of the existence of atoms. From now on, there could be no doubt about it: atoms were real.

And that unlikely chain of events is the story of how the atom was discovered – it all started on Robert Brown's expedition to Australia on board the *Investigator*.

Rutherford's atom

Ernest Rutherford was born in the rural South Island of New Zealand in 1871, the fourth of twelve children. He grew up on a farm, herding cows and riding horseback. It was this adventurous, pioneering spirit and enthusiasm for discovery which propelled Rutherford to become the greatest experimental physicist of his era. Rutherford has been called "the Newton of atomic physics".

Rutherford had an uncanny instinct for designing ingenious experiments using primitive equipment which revealed the structure of the atom. As Richard Rhodes says in his book *The Making of the Atomic Bomb*, Rutherford "won the atom. He found its constituent parts and named them. With string and sealing wax, he made the atom real."

Rutherford won a scholarship to Cambridge University to work in the Cavendish Laboratory under the leadership of J. J. Thomson. Thomson had been performing experiments with the recently-invented cathode ray tube. The cathode ray tube consists of an electrically negatively-charged heated filament called a *cathode*, facing a positively-charged *anode*. All the components are contained in a vacuum in a sealed glass tube:

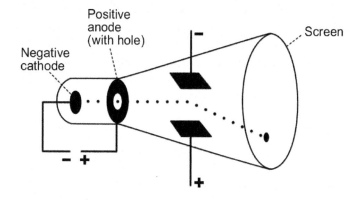

As the temperature of the cathode increased, it had been found that a stream of particles left the cathode, attracted to the positively-charged anode. If these particles hit a fluorescent screen, they made a small spot of light on the screen (this is how a cathode ray tube forms the main tube on old CRT televisions). Thomson discovered these particles could be deflected by electrically-charged plates, attracted to the positive plate, and their path could also be bent by a magnetic field. He named these particles *electrons*, the first atomic particle to be identified.

It was shown experimentally that the part of the atom which was left behind when these electrons were emitted was heavier, and was positively-charged. With this in mind, Thomson developed a model of the atom as a big lump of positively-charged material into which electrons were evenly distributed. This was called the "plum pudding" model (because the electrons were like raisins evenly distributed inside a pudding).

However, one of Rutherford's experiments soon revealed that this plum pudding model was incorrect.

By 1907, Rutherford had moved to Manchester University. In Manchester, Rutherford performed experiments involving recently-discovered radioactivity. He enlisted the help of Hans Geiger who had developed a device which could detect individual particles of radiation. Rutherford performed an experiment in which these particles (produced by radioactive radium) were fired at thin gold foils. He found some of the particles were deflected as they passed through the foil, but, more mysteriously, he found that some of the particles appeared not to get through the foil at all – they just vanished. Almost as an afterthought, Rutherford placed a detector in front of the foil instead of behind it, and he discovered – to his great surprise – that some of the particles were being reflected back from the foil.

Rutherford was amazed at this discovery: "It was quite the most incredible event that has ever happened to me in my life. It was almost as incredible as if you fired a 15-inch shell at a piece of tissue paper and it came back and hit you. On consideration I realised that this scattering backwards must be the result of a single collision, and when I made the calculations I saw that it was impossible to get anything of that order of magnitude unless you took a system in which the greatest part of the mass of the atom was concentrated in a minute nucleus."

This meant that Thomson's plum pudding model was wrong. Instead, it meant that atoms had a hard centre, a positively-charged *nucleus*.

Rutherford proposed a model of the atom which consisted of electrons orbiting a nucleus composed of positively-charged *protons*. The electrons were held in orbit due to their electrical attraction to the protons. This arrangement resembled a small Solar System, and was called the *Rutherford model* or *planetary model.*

However, the mass of atoms was about twice the mass which would be expected if the nucleus was composed of just protons. Rutherford proposed that the nucleus also contained other particles called *neutrons,* each neutron being approximately the same mass as a proton. These particles were called neutrons because they were electrically neutral. The existence of neutrons was confirmed in 1932.

Atoms are made out of these three particles: electrons, protons, and neutrons. As an example, the following diagram shows Rutherford's planetary model of a carbon atom. It is composed of six negatively-charged electrons orbiting a nucleus which is composed of six positively-charged protons and six neutrons:

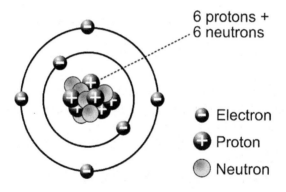

The death spiral of the electron

However, it was known that there was a problem with this planetary model. Just as in the case of a planet in the Solar System, eventually the orbit loses energy and the planet would crash into the Sun. A very similar scenario was predicted with this model of the atom. Newton had shown that any object moving in a circle is undergoing an acceleration (a deviation from its natural straight-line path), and according to Maxwell's theory of electromagnetism, any electrically-charged particle will lose energy in the form of radiation if it is accelerated. The electron would inevitably lose energy and crash into the nucleus in a thousandth of a billionth of a second.

Robert Oerter describes this rather disastrous scenario in his book about particle physics called *The Theory of Almost Everything*:

> *An electron in such an orbit would have to emit electromagnetic radiation, thereby losing energy, which would send it into a "death spiral", which could not end until the electron reached the nucleus. With all the negatively charged electrons in the nucleus cancelling out the positive nuclear charge, there would be no electric repulsion keeping the nuclei at atomic distances from each other. In a fraction of a second a house would collapse to the size of a grain of sand.*

So what was holding the electrons in their orbit? A solution was proposed by Rutherford's most eminent student, Niels Bohr.

In 1912, Bohr came to Manchester from Cambridge specifically to work under Rutherford. Bohr was concerned about the stability of Rutherford's atom. Rutherford had correctly discovered that electrons orbit the nucleus, but he provided no explanation as to what kept them in orbit. Bohr believed an answer was to be found from the new physics of quantum theory. Bohr was aware of the previous discovery of Max Planck that the energy radiated from a hot body was composed of "quantized" chunks of energy (described in my previous book). This seemed to indicate a new physics beyond classical physics. Bohr suggested that the energy of the orbiting electrons was not continuous, but was instead only allowed specific quantized values. This would result in electrons only being able to occupy specific "allowed" orbits. This restriction on electron energy would prevent the electrons from spiralling into the nucleus.

Bohr's theory resulted in a stable atom. Also, Bohr's theory predicted that radiation of certain specific energies would be emitted by the atom when an electron jumped between these allowed orbits. This prediction of specific energy levels perfectly matched the clearly-defined spectral lines emitted by heated materials.

So Bohr appeared to have found a solution. But it was a rather unsatisfactory solution in that it was a mix of the old classical physics (an atom was still being modelled as an electron spinning around the nucleus in just the same way that Newtonian physics predicted the orbit of planets around the Sun), and the new quantum theory.

A more elegant explanation was provided by the French physicist Louis de Broglie (pronounced "de Broy") in 1923. De Broglie was aware of Einstein's earlier explanation of the photoelectric effect (covered in my previous books) which suggested that a light wave was actually composed of particles called *photons*. Hence, there appeared to be a strange duality between waves and particles. De Broglie made the brave and inspired suggestion that this duality also worked in the other direction: that any particle also had a wavelike nature.

One consequence of this idea was that it provided an ingenious explanation of why electrons were held in their orbits around the nucleus. We find the solution if we no longer consider an electron as being an orbiting particle – like the Earth orbiting the Sun – but instead consider it as being a wave around the nucleus. We could imagine that only a whole number of wavelengths could be allowed around the nucleus: if this was not the case, the wave would not join up with itself when it returned to its starting point and it would interfere destructively with itself.

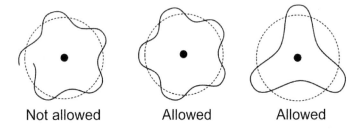

Not allowed　　　　**Allowed**　　　　**Allowed**

Different electron orbits represented a different number of whole wavelengths. Electrons could jump between orbits, but only by emitting or absorbing a photon of a fixed ("quantized") amount of energy. This model also implied that there would always be a minimum energy possessed by the electron (called the *ground state*) corresponding to the lowest frequency wave (analogous to the fundamental frequency of a wave on a string).

Hence, this wavelike model of the electron meant the death spiral of the electron into the nucleus could be avoided. And this is essentially the principle by which atoms achieve their size, and prevents houses collapsing into a grain of sand!

What is a particle?

Before we reach the end of this introductory chapter, there is a question which you might well be wondering. The question is: what are particles actually made from?

In order to answer this question, we have to examine particles at the highest level of detail. However, we certainly cannot use a conventional microscope to look at particles. That is because visible light has a wavelength between 4×10^{-7} metres and 7×10^{-7} metres, while atoms are much smaller, approximately 10^{-10} metres across. This is important, because it is not possible to see any object which is smaller than the wavelength of the light which illuminates the object. Hence, it is never going to be possible to see an atom with a conventional microscope.

However, the wavelength associated with a particle is inversely proportional to the energy of that particle. In other words, particles with more energy have shorter wavelengths (according to the previously-discussed theory of Louis de Broglie) and can therefore be used to probe smaller distances. As an example, the wavelength of an electron can be up to 100,000 times smaller than the wavelength of a photon (a particle of light), a principle which gave birth to electron microscopy.

In a modern particle accelerator, it is possible to achieve observations using particles with far higher energy than those used in an electron microscope. Using wavelengths as small as 10^{-16} metres it was possible to detect the existence of the smallest known particles. However, as accelerator energies have increased, with wavelengths now down as small as 10^{-18} metres, no smaller structure has been found in these particles. It would appear that these are *elementary particles* in that they are truly fundamental having no internal structure – no constituent parts. It is simply not possible to break them down into smaller units.

So how big is an elementary particle? As the energy of particle accelerators has increased, these particles appear to be smaller than anything we can detect. In fact, it is believed that these particles have no actual size: they should be considered as infinitely small pointlike particles. It might be said that they have no *spatial extent* – they do not extend into space. Essentially, they are not made of anything! Which makes a kind of sense: if they were made of any material, it would be possible to split that material into two parts to create smaller particles. In which case, the original particles would not be elementary (as they could be split into two "more fundamental" particles).

By this logical argument, it would appear that a truly elementary particle would **have** to be pointlike, or else it could not be considered to be truly elementary.

This type of behaviour, and this strange type of argument, might appear completely bizarre and counter-intuitive. How on earth is it possible to have objects which cannot be broken into smaller objects? How on earth is it possible to have objects which have no spatial extent? The problem is that we are only used to operating in the human-scale ("macroscopic") world, and things work very differently at that vastly larger scale: all objects can be broken into smaller objects, all of these composite objects take up volume in space, etc.

However, at the fundamental scale of elementary particles, the world works very differently, and in very counter-intuitive ways. Another counter-intuitive fact about elementary particles is that all particles of a particular type are identical. For example, all electrons are exactly the same, as if they are perfect copies (we will discover the implications of this later in the book when we consider the Pauli exclusion principle). Again, this has no analogy in our macroscopic world. In our macroscopic world, we always have the option of "delving deeper", to sub-analyse an object to discover distinguishing details. For example, if we want to distinguish two identical models of car, we might divide the car into its constituent parts, and then consider the cars' number plates. Hence, all macroscopic objects can be distinguished. However, when we are dealing with elementary particles, we no longer have access to any lower distinguishing level. This is a clear sign that we are dealing with the lowest level of reality.

When dealing with these fundamental concepts, we have to make the effort to break away from our human-scale, macroscopic preconceptions about how things behave, and how objects are structured. Because only then will we have a chance of understanding the elementary structure of Nature.

This book will consider the structure of the atom, and is essentially divided into three parts. The first part might be described as a study of "the forces which hold the atom apart" (in this chapter we have already seen how energy quantization prevents the electrons from spiralling into the nucleus). The second part of the book might be described as a study of "the forces which pull the atom together", in which we will examine the fundamental forces which hold the electrons and the atomic nucleus together. And the third part of the book considers the balance which arises between these two opposing forces: the forces which hold the atom apart, and the forces which pull the atom together.

It is by considering this concept of "balance" in this third part of the book that we will find a remarkable similarity between the forces which hold the atom together and the forces which hold the universe together. We will discover that the very large and the very small have much in common.

2

ADVENTURES IN THE MATRIX

When Ernest Rutherford gave an interview to the *London Daily Herald* in 1933, he tried to explain how small the atom was:

If everyone in the world spent twelve hours a day placing individual atoms into a single thimble, a century would elapse before it was filled.

At first glance, the intended message of Rutherford's quote appears clear: atoms are inconceivably small. However, there is another possible interpretation because, of course, all sizes are relative. The alternative interpretation is that humans (and thimbles, for that matter) are inconceivably huge.

In similar fashion, whenever I hear the usual quote that humans are insignificantly small in relation to the size of the universe, I get mildly irritated. Because, of course, on the atomic scale (which is a much more sensible standard scale of size) human beings are cosmos-straddling giants. A typical human contains 7,000,000,000,000,000,000,000,000,000 atoms. We are certainly not "insignificantly small" by that definition.

In fact, it is this extraordinarily huge size of human beings (and our experimental apparatus) which causes us such great difficulty as we attempt to explore the atomic world, because much of the true behaviour of atoms becomes hidden from our eyes. Instead, all we ever consider is phenomena in which the individual behaviour of atoms is "averaged out". Most notably, the peculiar quantum mechanical behaviour of individual atoms is obscured as we only ever consider systems which are composed of billions of atoms. However, if we really want to obtain an accurate picture of the behaviour of individual atoms, it is precisely this quantum mechanical behaviour which we need to capture.

And so, this chapter presents a brief introduction to quantum mechanics, describing the actual techniques used by physicists in order to construct an accurate model of the atom.

Basic matrix operations

Central to this discussion of quantum mechanics will be the mathematical object known as a *matrix*. You might have been introduced to matrices ("matrices" is the plural of "matrix") in your mathematics class in school, and you might have suspected that you would never use such an obscure piece of mathematics in your everyday life. Well, unless you found employment in some fields of science and engineering then you might have been correct. However, in this book you will finally get a chance to use your skills as you discover just how useful matrices really are.

The discussion will show why matrices are so central to quantum mechanics, and why solving problems in quantum mechanics involves solving a problem described by matrices. You will get much more out of the rest of the book if you ensure you understand the concepts in this section. I will be asking you to perform some important matrix calculations later in this book. You won't believe what we can achieve!

A matrix is a square (or rectangular) grid of numbers. Here is an example of a 2×2 matrix (it has two rows and two columns):

$$\begin{bmatrix} 2 & 4 \\ 7 & 3 \end{bmatrix}$$

We might ask how matrix multiplication (the multiplication of two matrices) might be achieved. For example, how could we multiply the following two matrices:

$$\begin{bmatrix} 2 & 4 \\ 7 & 3 \end{bmatrix} \times \begin{bmatrix} 5 & 6 \\ 1 & 5 \end{bmatrix}$$

We can multiply these two matrices by the following sequence of steps.

First, consider the first row of the first matrix and the first column of the second matrix (shown in the two dashed ellipses in the following diagram). Take the first number in the first row of the first matrix (which is 2) and multiply it by the first number in the first column of the second matrix (which is 5). This gives us our first intermediate result of 10 (we must remember this). Then take the second number in the first row of the first matrix (which is 4) and multiply it by the second number in the first column of the second matrix (which is 1). This gives us our second intermediate result of 4.

We add our two intermediate results (10 and 4) together to get the answer 14, and that number goes in the first position in our result matrix:

$$\begin{bmatrix} 2 & 4 \\ 7 & 3 \end{bmatrix} \times \begin{bmatrix} 5 & 6 \\ 1 & 5 \end{bmatrix} = \begin{bmatrix} 14 & \\ & \end{bmatrix}$$

To calculate the next entry in our answer matrix, once again we have to consider the first row of our first matrix, but now we consider the second column of the second matrix. We perform a similar series of steps to the first case, but this time we need to calculate $(2 \times 6) + (4 \times 5)$ to give an answer of 32:

$$\begin{bmatrix} 2 & 4 \\ 7 & 3 \end{bmatrix} \times \begin{bmatrix} 5 & 6 \\ 1 & 5 \end{bmatrix} = \begin{bmatrix} 14 & 32 \end{bmatrix}$$

We continue in a similar fashion until we have filled all the positions in our result matrix. So here is the answer of our matrix multiplication:

$$\begin{bmatrix} 2 & 4 \\ 7 & 3 \end{bmatrix} \times \begin{bmatrix} 5 & 6 \\ 1 & 5 \end{bmatrix} = \begin{bmatrix} 14 & 32 \\ 38 & 57 \end{bmatrix}$$

You might want to check the steps yourself to make sure it is the correct answer. $(7 \times 5) + (3 \times 1) = 38$ and $(7 \times 6) + (3 \times 5) = 57$.

One vitally important feature about matrix multiplication is that it is *noncommutative*. Conventional multiplication is commutative in that it does not matter in which order the numbers are multiplied, the answer will be the same. For example, 7×5 will give the same result as 5×7. However, matrix multiplication is noncommutative in that the ordering does matter. If the ordering of the matrices is reversed, the answer will be different. You might want to check this by considering the previous example, but this time placing the second matrix before the first matrix. When the multiplication is performed, you should get a different answer:

$$\begin{bmatrix} 5 & 6 \\ 1 & 5 \end{bmatrix} \times \begin{bmatrix} 2 & 4 \\ 7 & 3 \end{bmatrix} = \begin{bmatrix} 52 & 38 \\ 37 & 19 \end{bmatrix}$$

So this is something very important which I want you to remember as we will be returning to this point several times in this book: **matrix multiplication is noncommutative**. You get a different result for your multiplication when you reverse the ordering of the matrices.

Vectors

The mathematical construction known as a *vector* is closely associated with matrices. A vector is a line of a specified length which has a definite direction – essentially it is an arrow. In three-dimensional space, we could imagine a vector extending outwards from the origin of our coordinate system, as is the case with the dashed arrow in the following diagram:

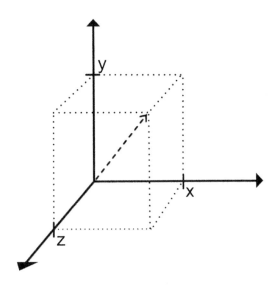

You will see that the dashed arrow (the vector) points to a point in space which is a distance x along the horizontal axis, and is a distance y along the vertical axis, and a distance z along the axis which is coming out of the page. You can see that this vector could actually point to any position in space, as long as we have the (x, y, z) coordinates of the position. Hence, three coordinates are enough to define this vector, and we could list those coordinates in matrix form as a matrix which has only one column:

$$\begin{bmatrix} x \\ y \\ z \end{bmatrix}$$

A vector is therefore essentially just a matrix, and so we can perform matrix multiplication by multiplying a vector by a matrix. As we shall see, the result of the multiplication will be another vector. Hence, a matrix can be used for **transforming** a vector into another vector.

Here is an example of a matrix multiplication being used to transform a vector which has just two coordinates [-2, 4] (as an example, a vector drawn on a flat sheet of paper would only have two coordinates: x and y):

$$\begin{bmatrix} 0 & 1 \\ -1 & 0 \end{bmatrix} \times \begin{bmatrix} -2 \\ 4 \end{bmatrix} = \begin{bmatrix} 4 \\ 2 \end{bmatrix}$$

The result vector is calculated in the usual method. The first entry is $(0 \times -2) + (1 \times 4)$ which is 4, and that goes in the first position in the result vector. The second entry would be $(-1 \times -2) + (0 \times 4) = 2$.

So what is the result of this transformation? A vector with the coordinates [-2, 4] has been transformed into a vector [4, 2]. Let us plot these two positions on a graph:

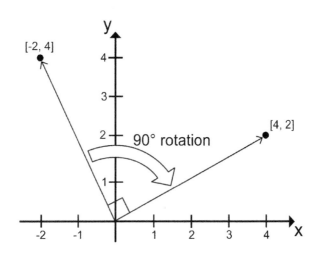

You will see from the graph that the result of this matrix multiplication is to rotate the vector about the origin by 90 degrees.

This ability of matrices to perform transformations is an incredibly useful property. As we shall now discover, this is the reason matrices are used in quantum mechanics.

The quantum measurement

We only have to consider the implications of quantum mechanics when we are dealing with the smallest parts of Nature, which are particles. True, the rules which apply to particles inevitably apply to larger objects as well, but the counter-intuitive behaviour of quantum mechanics gets "averaged out" as we move up the scale from individual particles to macroscopic ("human scale") objects composed of trillions of particles. We simply don't have to deal with the quantum rules directly in our everyday lives.

However, in this book we are interested in the behaviour of particles, so a good understanding of quantum mechanical behaviour is going to be crucial.

I considered the logic of quantum mechanics in my first book, hopefully trying to convey the impression that there is a clear and consistent underlying logic to quantum mechanical behaviour (even though the "weirdness" of quantum mechanics is often stressed elsewhere). It was stated how quantum mechanics is important whenever we want to extract some information about the state of a system. That is just a grand way of saying that quantum mechanics is all about measurement. When we talk about "measurement" you might have an impression of someone using a tape measure, but, really, measurement is a much more general concept. Whenever we detect an object, for example, we are measuring the position of the object and determining that it resides in a particular location. In fact, in the most general case, we might consider "measurement" to be **any** interaction between particles (because that interaction has the ability to pin down the position of a particle).

In my first book, I described the principles of measurement at the quantum scale in very simple terms:

1) Before we measure certain properties of a particle (e.g., position, momentum) the particle behaves as though it has **all possible values** for that property.

2) After we have performed the measurement, we find the particle property takes only one of the possible values at random. It is fundamentally impossible to predict which value the property will take.

And those two simple principles, basically, are the underlying principles of quantum measurement. Hence, they are the underlying principles of quantum mechanics itself. It is really not so difficult or weird!

23

With these two principles in mind, I presented a useful analogy to quantum measurement: a roulette wheel. Before the measurement is taken, we might imagine the ball spinning around the wheel to be in all possible states, certainly it has the potential to take all possible states. In this pre-measurement phase, we might consider the system to be in a *superposition* state (i.e., all possible states). However, when the measurement is taken, the ball occupies only one clearly-defined state. The number of the slot indicates the value of the measurement. It is fundamentally impossible to predict which slot the ball will end up in. In other words, it is fundamentally impossible to predict the outcome of a measurement in advance: quantum mechanics can only give you the probability of a certain outcome.

As an example of these principles in practice, my first book described the famous *double-slit experiment*. According to Richard Feynman: "The double-slit experiment has in it the heart of quantum mechanics. In reality, it contains the **only** mystery of quantum mechanics."

In the double-slit experiment, a light source is in front of a board. Two narrow slits are cut into the board. Light can only pass through these two slits, and the light which passes through the two slits illuminates a screen behind the board. The two light rays from the two slits meet at the screen. Due to the wavelike nature of light – and the resultant constructive and destructive interference – this results in a characteristic pattern of dark and light bands projected onto the screen (for details, see my first book).

This all makes sense, and is in agreement with the classical (i.e., non-quantum) view of waves. Indeed, this effect can even been seen in water waves passing through two slits.

However, the real surprise happens when the intensity of the light is reduced to such an extent that only one particle (one photon) is emitted at a time. In other words, at any point in time there is only a single photon travelling from the light source, through one of the two slits (we do not know which slit), before hitting the screen behind the board. Over time, these individual photons build up a pattern on the screen. Amazingly, the pattern on the screen still exhibits the dark and light interference bands. It appears that the photon is interfering with itself! In fact, **it appears that the photon is passing through both slits at once**.

As amazing as this result might appear, it is in line with the simple principles of quantum mechanics I described earlier in this section. Before we make a measurement of a particle's position (i.e., before we localise the particle by making it hit a screen) the particle behaves as if it has all possible values, as if it is passing through both slits at the same time. At this point, the particle is in a superposition state of all possible values.

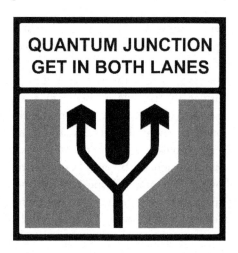

But, you might argue, one particle is only making one dot on the screen. Surely this means that the particle is truly only going through one slot? If we could install detectors on each slit, surely we might be able to detect which slit the particle is actually passing through? If we try this approach, true enough, we discover the particle appearing to only travel through one slot – **but this makes the interference pattern on the screen disappear!** By detecting the particle, we are effectively localising it, and forcing it to appear in only one position. The act of measurement takes the particle out of its superposition state. Again, this agrees with the principles of quantum mechanics I described earlier this section: after we have performed the measurement, we find the particle property takes only one of the possible values at random.

At this point, we start to realise why all our work on matrices and vectors was worthwhile. In order to see why that is the case, we have to realise that a particle in a superposition state is really in a combination of all possible states. And we can represent a combination by a vector.

In order to see how we can do this, let's expand the double-slit experiment into a triple-slit experiment. In other words, we get our knife out and cut another slit in the board so that the particle can now pass through three slits in the board instead of just two slits.

Before the particle hits the screen and is detected, we have to consider the particle as being in a superposition state, which means we now have to consider the particle is passing through all three slits. The state of the particle is therefore in a combination of three states: "Going through slit A", "Going through slit B", and "Going through slit C". In the following diagram, I hope you can see how this superposition state of the particle can be represented as a vector (the dashed arrow) which is composed of all three possible states:

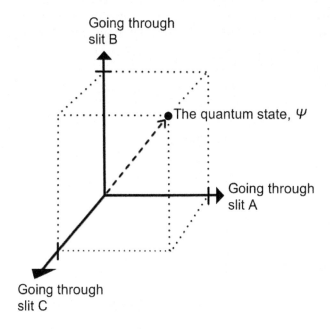

Going through
slit B

The quantum state, Ψ

Going through
slit A

Going through
slit C

You will notice that this superposition state of the particle is referred to as the *quantum state*, and is denoted by the Greek letter Ψ (pronounced "sigh"). This is the standard notation used to denote the quantum state of a particle.

So this explains how vectors are a perfect mathematical tool for representing quantum states. Please note that the vectors in the previous diagram no longer represent the three coordinate axes of space. No, the three vectors are now purely a mathematical abstraction representing the three possible states of the particle. This type of vector space which is used for representing quantum states is called a *Hilbert space* (after the great German mathematician David Hilbert).

So what happens when we make a measurement? Well, the particle is pulled out of its superposition (multi-valued) state into a single well-defined state. For example, when we try to detect the position of the particle in our triple-slit experiment, we find it only passing through one slit. As you will see in the following diagram, in state vector terms this is represented by a rotation of the quantum state vector from its previous superposition state to one of the well-defined single states (in this case, going through slit B):

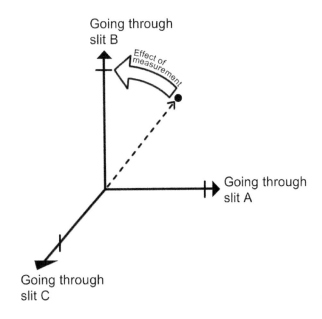

And what have we already seen are very useful tools for rotating vectors? Yes, matrices! (Remember earlier in this chapter we showed how matrix multiplication can rotate a vector). So all our work with matrices was not wasted. We will now see that matrices are the key tool for analysing what happens when we make a quantum measurement.

However, the way we use matrices to predict the possible outcomes of measurements is rather different from this rotational model. The next section describes the procedure which really lies at the heart of all quantum mechanical measurements.

You just need to know the magic word …

The magic word: "Eigen"

You will find in quantum mechanics that a particular word crops up time and time again. It is a German word, emphasising that Germany led the world in the development of quantum mechanics in the early decades of the 20th century. The word is "eigen", and – in the German style – you will find the word added before another word to make a compound word, such as "eigenstate", "eigenvector", or "eigenvalue".

Apparently in German, the correct translation of "eigen" is "personal" or "idiosyncratic". However, when I see the word "eigen" I think of only one word, and that word is "allowed".

As an example, an *eigenvalue* is the value you will measure when you perform a measurement or observation on a quantum system. Only a few, certain values can possibly result from your measurement: you will find you will only measure one of a few "allowed" values.

As an example, in the previously-described triple-slit experiment, when we perform a quantum measurement on the particle (i.e., measuring its position to detect which slit it went through) we will obtain one of only three possible values: slit A, slit B, or slit C. In other words, only three possible values are allowed. These would represent the eigenvalues of the system.

Remember the analogy I presented earlier of a ball spinning round a roulette wheel? I said that before the measurement is taken, the ball spinning around the wheel represents the multi-valued superposition state. However, after the measurement is taken, the ball only occupies one clearly-defined slot. The numbers on these clearly defined slots represent the allowed eigenvalues we will measure.

So how do we calculate our possible eigenvalues? Firstly, we have to decide what property we want to measure, because different properties have different eigenvalues. Each different measureable property value (called an *observable*) is associated with a different matrix operation. And that particular matrix operation is achieved by using a particular *operator matrix* (sometimes simply called an *operator*). For example, if you want to know the possible energies which a particle can have, you would use the energy operator matrix. Or, if you want to know the possible values for a particle's momentum, you would use the momentum operator matrix.

This use of a particular operator in order to produce a measureable value emphasises the importance of the observer in quantum mechanics. It is as though the very act of observation creates the measured value. Before the operator is applied, it is meaningless to talk of the value of a property of a particle.

So, when we have selected a suitable operator matrix, what happens next? Well, it turns out that the operator matrix contains within it all the information we need, but we have to extract that information in a rather unusual manner. Remember that the possible states of the particle (e.g., "slit A", "slit B", etc.) can be represented by vectors in a vector space. These vectors are called *eigenvectors* (i.e., the "allowed" states after measurement). So how are eigenvectors and eigenvalues related? The relationship is quite peculiar, but it is vitally important: **The result of multiplying an eigenvector by the operator matrix results in the same eigenvector – but multiplied by a number. And that number happens to be the eigenvalue we measure.**

This method is represented in the following diagram:

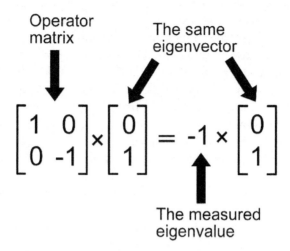

Take some time to ensure you understand the previous diagram. This eigenvalue/eigenvector approach is the **absolutely central** method for analysing quantum mechanical measurements. An actual example will be presented later in this chapter.

For a particular operator matrix, only a few very specific eigenvectors can successfully satisfy this requirement. So a particular operator matrix defines a set of possible eigenvectors. And, as each eigenvector is associated with just one particular measureable eigenvalue, this means that a particular operator matrix gives us the complete set of possible measurements we will find. As Leonard Susskind and Art Friedman say in their book *Quantum Mechanics*: "An operator is a way of packaging up states along with their eigenvalues, which are the possible results of measuring those states."

Crucially, the noncommutative property of matrices (which we considered earlier in this chapter) is known to have important consequences for the ordering in which measurements are taken in quantum mechanics. The

Heisenberg uncertainty principle is a consequence because the order in which you take measurements is important: if you measure a particle's momentum first, and then its position, you will get a different result if you performed the measurements in the opposite order. This is due to matrix multiplication – which lies at the heart of quantum mechanical measurements – being noncommutative: the order does matter.

So this is a very mathematical approach to producing "allowed" eigenvalues. But is there a more intuitive approach? Well, yes there is. Where previously have we encountered "allowed" values before? Remember back to the work of Louis de Broglie in the previous chapter who suggested that only an integer (whole number) of wavelengths could be allowed in the orbit of an electron around the nucleus. This meant that the energy of the electron could not be continuously variable, but had to be divided into discrete chunks. In other words, only certain energy levels were allowed. Well, these allowed values of energy precisely correspond to the allowed eigenvalues of energy for the electron.

It has been shown that the eigenvalue approach to finding allowed values, and the "whole number of wavelengths" approach to finding allowed values are mathematically equivalent. In fact, in 1926 Erwin Schrödinger first presented his famous Schrödinger wave equation in a paper entitled *Quantization as an Eigenvalue Problem*.

The following diagram shows how allowed energy states are related to the number of whole wavelengths:

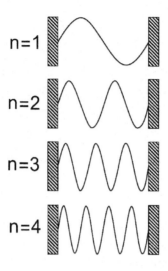

So next time you might be having difficulty in imagining the "allowed" eigenvalues produced by a matrix calculation, just imagine the problem in terms of an allowed number of integer wavelengths instead.

This wavelike nature of the quantum state means the quantum state (or quantum state vector) is frequently referred to as the *wavefunction*. Personally, I generally find it easier to think in terms of a quantum state as a vector, and that is the approach I have followed in this book. But just bear in mind that the two terms both refer to the same thing.

Certainly, this wave-based approach hopefully gives you more of a feel as to how discrete eigenvalues can emerge as the only allowed values for a property. And it is these discrete (rather than continuous) values which we find at the lowest level of Nature, and it is these "chunks" – the *quanta* – which give quantum theory its name.

Spin

In 1921, Otto Stern and Walther Gerlach performed an experiment in Frankfurt which revealed a new property of particles. The experiment is now known as the *Stern-Gerlach experiment*. In the experiment, Stern and Gerlach sent a beam of particles through a magnetic field. They observed that the particles were deflected, and left two distinct marks on a screen. In order to explain this, it was proposed that the particle might be spinning about an axis. After all, a moving charge generates a magnetic field, so this would result in the particle behaving as if it was a bar magnet. If the particles turned into bar magnets, then this would explain the observed magnetic deflection.

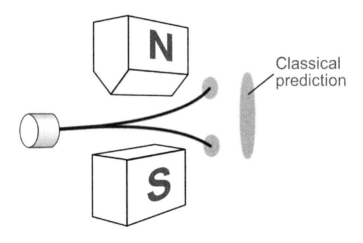

Classical prediction

However, there was a problem with this picture of a spinning particle. If we imagine the particles as spinning around an axis, then we would imagine this axis to be randomly-oriented when the particles were emitted from the gun. We would certainly expect the effect of the magnetic field to deflect the particles, but we would expect this to result in a smooth smearing of the particles when they hit the screen (see the "classical prediction" in the previous diagram). However, this is not what happens. When the particles hit the screen, they cluster into two clearly-defined groups.

This clustering into two groups suggests that when we measure the spin of a particle along a particular axis (such as north-south in the previous example) then only two results are possible. Imagine the curled fingers of your right hand representing the direction of spin of the particle:

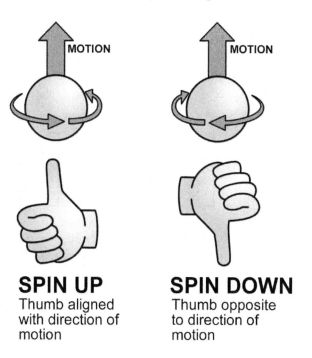

SPIN UP
Thumb aligned with direction of motion

SPIN DOWN
Thumb opposite to direction of motion

As you can see from the previous diagram, we can understand the two possible states of the particle by using our right hand with our fingers curled up and our thumb pointed out. In this way, the curling of the fingers forms a *spin vector* representing the rotation of the particle. As you can see from the diagram, if our thumb is directed in the direction of motion of the particle then that represents "spin up". Alternatively, if our thumb points in the opposite direction to the direction of motion, that represents "spin down". When we measure the particle, these are the only two possibilities we will encounter.

So this reminds us of eigenvalues: only two values are **allowed**: spin up or spin down. It is as if our quantum roulette wheel has only two slots into which the particle can land. So we come to the conclusion that the spin of a particle is an example of quantum mechanical behaviour.

So if spin is quantum mechanical in nature, let us use our knowledge of eigenvalues and eigenvectors to express spin in a useful mathematical form.

Firstly, we can see that there are two possible states of spin: "spin up" and "spin down". These will be our two "allowed" eigenvector states, and we can represent these by the two vectors:

$$\text{Spin up:} \begin{bmatrix} 1 \\ 0 \end{bmatrix} \quad \text{Spin down:} \begin{bmatrix} 0 \\ 1 \end{bmatrix}$$

Now, what about the value we actually measure when we make a measurement? In other words, what values would we expect for our eigenvalues? Logically, we would expect to measure a value of +1 for spin up, and -1 for spin down.

And that is all the information we need. We are now in a position to derive the all-important operator matrix which describes the spin of a particle. Let us represent the operator matrix we need to find by:

$$\begin{bmatrix} a & b \\ c & d \end{bmatrix}$$

We are now left with the task of finding the values of a, b, c, and d. How do we do that? Well, using our knowledge of eigenvalues and eigenvectors, we know that the operator matrix must satisfy the two conditions for "spin up" and "spin down":

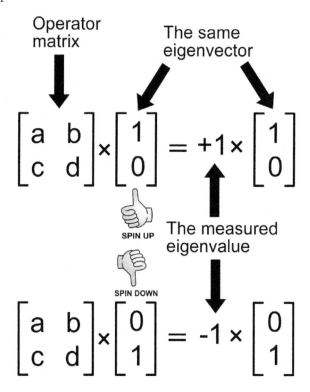

These two conditions can only be satisfied if $a=1$, $b=0$, $c=0$, and $d=-1$. Hence, the operator matrix which describes spin is:

$$\begin{bmatrix} 1 & 0 \\ 0 & -1 \end{bmatrix}$$

(You might want to check that this operator matrix satisfies both of the two previous eigenvalue/eigenvector operations.)

So this is great! This single operator matrix contains all the information we need to describe both of the possible spin states about an axis, and the spin values we can measure.

However, there are two more axes about which we could measure spin, and the operator matrices for these additional axes can be found in a similar fashion.[1] The set of the three operator matrices which describe spin about the three axes (x, y, and z) are called the *Pauli spin matrices* (named after Wolfgang Pauli):

---

[1] For details, see Section 3.4 of Leonard Susskind and Art Friedman's book *Quantum Mechanics*.

The Pauli Spin Matrices

$$\begin{bmatrix} 0 & 1 \\ 1 & 0 \end{bmatrix} \begin{bmatrix} 0 & -i \\ i & 0 \end{bmatrix} \begin{bmatrix} 1 & 0 \\ 0 & -1 \end{bmatrix}$$

where i is the square root of -1 (if you multiply i with i you get -1).

So that is very satisfying. From our knowledge of quantum mechanics we have been able to derive the matrices which are used to describe the behaviour of particle spin, and we will be encountering these Pauli spin matrices again later in this book.

However, we produced these matrices merely by observing the behaviour of particles. It would be so much better (and rather amazing) if we could derive these matrices from basic principles.

Well, that is precisely what we shall do in the next chapter.

But … what is spin?

We have covered a lot of important ground in this chapter, and at quite a pace. But we have managed to get to this point without ever specifying what, precisely, is meant by the spin of a particle. According to most popular science books, there is apparently little doubt that particle spin has nothing to do with actual physical rotation of the particle.

Well, sorry, I can't accept that. I can't help thinking that if you repeat a statement often enough then eventually it becomes a kind of fact which gets copied and repeated and nobody ever gets round to checking it, so it takes on a life of its own.

It is time for a correction.

In the Stern-Gerlach experiment described earlier this chapter, you will remember that particles deflected in a magnetic field behaved **precisely** as if they were actually physically spinning, the spinning particles turning into miniature bar magnets. This was the initial motivation for suggesting that particles have spin and, indeed, provided the motivation for the term "spin". Indeed, no one disputes that these particles possess the property of angular momentum. So why do so many authors appear to suggest that there is no physical rotation? It appears that this is because spin does not behave in the same way as the classical notion of spin with which we are all acquainted. When a measurement is made of quantum spin, the angular momentum of a particle is quantized into chunks of Planck's constant, h. As we have seen, we interpret this as either "spin up" or "spin down" for the particle, which is obviously not the case for a classically spinning object.

So on the basis of these quantum peculiarities, should we conclude that the particle is not actually physically spinning, or should we instead conclude that this is precisely the sort of behaviour we should expect in quantum mechanics?

I firmly believe we should come to the latter conclusion.

Firstly, when we measure the spin of a particle, some might be surprised that the act of observation appears to force the particle to choose between either "spin up" or "spin down". But this type of observer effect is found throughout quantum mechanics, and we should surely not be surprised by it. In fact, it would have been much more surprising if we had **not** found an observer effect! It would have been much more surprising if had found particle spin operating according to classical rules when we moved down to the quantum level.

Admittedly, particle spin is a very peculiar property. For example, in the next chapter we will be discovering that you need to spin an electron around twice to return it to its starting position! Crucially, though, these peculiarities do not necessarily signify that the particle is not physically rotating.

As was explained at the start of this chapter, when billions of particles work together to form macroscopic phenomena, there is an "averaging out" of quantum mechanical behaviour, and all humans ever see is the classical form of spin. Hence, quantum spin gets turned into classical spin by this averaging process. As in all examples of quantum mechanical behaviour, the classical behaviour arises from the quantum behaviour. But at the base level, they both arise from the same root cause: quantum spin becomes classical spin.

An accurate explanation of spin is provided by Robert Oerter in his book *The Theory of Almost Everything*:

> *What is this thing called spin? The electron's spin, as the name implies, has to do with rotation. Take a beam of electrons that are all spinning in the same direction and fire it at, say, a brick. If you could keep this up for long enough, and if there were no other forces acting on the brick, the electrons would transfer their rotation to the brick and it would begin to rotate.*

Thank you so much Mr. Oerter! It is reassuring to find a popular science author who is not content to repeat the usual "non-rotation" script.

A similar argument is presented by Luboš Motl in a blog posting entitled "The electron is spinning, after all" (http://tinyurl.com/electronisspinning).

And so that concludes our introduction to quantum mechanics and quantum spin. In the next chapter we will discover how one of the greatest mathematical physicists of all time used this knowledge of quantum mechanics and spin to obtain extraordinary insights into the structure of matter and the atom.

3

THE AMAZING MR. DIRAC

Paul Dirac was born in Bristol, England in 1902. His father was a harsh disciplinarian who imposed a strict work ethic on the young boy. As his mother later said: "His father's motto has always been to work, work, work, and if the boy had showed any other tendencies, then they would have been stifled. But that was not necessary. The boy was not interested in anything else." As a result of this sheltered upbringing, Dirac grew up quiet and withdrawn, and rather emotionally stunted. "This balancing on the dizzying path between genius and madness is awful", Einstein later said of Dirac. However, Dirac's talent, hard work and phenomenal mathematical ability eventually resulted in Stephen Hawking describing him as "probably the greatest British theoretical physicist since Newton."

Dirac's technical ability at mathematics was apparent from an early age. In 1919 as a young student, Dirac was profoundly affected by a breakthrough in the field of physics. As Graham Farmelo recounts in his biography of Dirac: "No event in Dirac's working life ever affected him as deeply as the moment when relativity 'burst upon the world, with a tremendous impact' as he remembered nearly sixty years later."

The predictions of special relativity agreed with Newtonian physics for speeds which were much lower than the speed of light, but special relativity modified Newton's laws for speeds close to the speed of light. Purely as a hobby while he was studying as an undergraduate in college in Bristol, Dirac practised upgrading Newtonian theories to incorporate special relativity. According to Dirac: "There was a sort of general problem one could take, whenever one saw a bit of physics expressed in a non-relativistic form, to transcribe it to make it fit in with special relativity. It was rather like a game, which I indulged in at every opportunity."

When he won a scholarship to Cambridge University, Dirac applied his mathematical virtuosity to quantum mechanics, operating at a far higher mathematical level than his fellow physicists who often came from an engineering background, and therefore viewed the atom as a mechanical object. Dirac's approach proved fruitful, showing that a purely mathematical approach was more suited to capturing quantum mechanical behaviour.

The Dirac equation

The challenge for physicists in the 1920s was to combine quantum mechanics with the special theory of relativity. Erwin Schrödinger had derived a famous equation which described the quantum mechanical behaviour of a particle, but the equation was not relativistic, which meant it did not describe the behaviour accurately as the particle's speed approached the speed of light. Paul Dirac was determined to find a relativistic equation which accurately described the quantum mechanical behaviour of an electron.

In Chapter Eleven of his biography of Dirac, Graham Farmelo describes the motivation and thought processes of Dirac in detail which provides a fascinating insight into Dirac's approach as he successfully derived one of the most important equations in physics, now simply known as the *Dirac equation*.

We will now derive the Dirac equation. What follows might appear to be rather complex. However, there is a considerable reward for us if we take our time and work through this. We will uncover the truth as to why the electron has spin. We will discover why the behaviour of matter particles (such as electrons) makes a solid object feel hard to the touch (and why, conversely, you cannot feel a ray of light). And we will discover how this leads to an understanding of all of chemistry: why the elements have their different characteristics, and how they combine to produce everyday materials. We will even discover why there is matter and antimatter.

And we will also be obtaining an insight into how one of the greatest minds in physics derived one of the most important results in the history of physics.

Dirac took as his starting point the most important equation in special relativity, which is the *energy-momentum relation*:

$$E^2 = (pc)^2 + (mc^2)^2$$

where E is the total energy of a particle, m is its mass, p is its momentum, and c is the speed of light (this equation was derived and considered in detail in my third book). This equation reveals the relationship between energy, mass, and momentum, and it applies to absolutely everything in the universe, from a rolling billiard ball to the energy and momentum of light itself.

You will see that if a particle is stationary, i.e., if its momentum, p, is set to zero, then the equation reduces to Einstein's famous equation $E=mc^2$. However, the energy-momentum relation is more general than Einstein's equation as it can be shown to apply even to massless objects, such as photons.

You will see that the energy-momentum relation can also be written as:

$$E = c\sqrt{p^2 + (mc)^2}$$

Dirac considered this equation and did not like the fact that the p and mc terms were squared. In his biography of Dirac, Graham Farmelo explained Dirac's motivation: "Believing that the relativistic equation would be fundamentally simple, he thought it most likely that the equation would feature the electron's energy and momentum just as themselves, not in complicated expressions such as the square root of energy or momentum squared."

So Dirac considered the content of the square root and tried to express it purely in terms of p and mc, without those terms being squared. What he then had to do was work out the relative amounts of each term.

So Dirac tried to represent the energy-momentum relation as two bracketed terms multiplied together:[2]

$$p^2 + (mc)^2 = (ap + bmc)(ap + bmc)$$

where a and b are some numbers which have to be calculated. They represent the amounts of p and mc which are required.

If we multiply-out the two bracketed terms, we find we get something which looks very close to the energy-momentum relation, but is not quite exactly right. The result includes a term which does not appear to fit (I call this the "unwanted part" in the following diagram):

$$(ap + bmc)(ap + bmc) =$$
$$ap(ap + bmc) + bmc(ap + bmc) =$$

This part closely resembles the energy-momentum relation

$$\boxed{a^2 p^2 + b^2 (mc)^2} + \boxed{(ab + ba) pmc}$$

The "unwanted part"

---

[2] As this is now the square of two identical terms, the square root of this expression is just one of these two terms. So if we take just one of these terms we will be able to eliminate the square root from the energy-momentum relation.

Considering the first part of the result, we can see that if we set a=1 and b=1 then the a and b terms effectively disappear (multiplying by 1 is the equivalent of not multiplying at all) and we are left with precisely the old energy-momentum relation. So it would appear that those two values for a and b would represent a solution to Dirac's relativistic equation for the electron. However, the problem is that the "unwanted part" does not vanish with those values of a and b.

So Dirac was stuck with a problem. He needed to find values for a and b such that a^2=1 and b^2=1. However you will see from the "unwanted part" that he also needed $(a \times b + b \times a)$ to equal zero to remove the entire "unwanted part". Unfortunately, it was clear that no simple numbers satisfied these requirements. In fact, bizarrely, the requirement for $(a \times b + b \times a)$ to equal zero appeared to indicate that a and b had to be mathematical objects for which a different order of multiplication produced a different result (so that $a \times b$ might produce a positive number, while $b \times a$ produced an equal negative number – the sum of those two numbers then being zero).

So where have we heard of this type of noncommutative multiplication before, multiplication in which the ordering matters?

The story is that Dirac was staring into the fireplace at Cambridge when he was struck by a stroke of genius. He realised that a and b were not simple numbers at all. Instead, he realised that matrices exhibited precisely this type of noncommutative multiplication which was required (as we discussed in the previous chapter).

Here is an example of two matrices which would meet Dirac's requirement:[3]

$$a = \begin{bmatrix} 0 & 1 \\ 1 & 0 \end{bmatrix} \quad b = \begin{bmatrix} 1 & 0 \\ 0 & -1 \end{bmatrix}$$

Remember, the first requirement is that the square of either of the matrices must be equal to one. So let's pick one of the matrices – say, matrix a – and square it (multiply it by itself):

$$\begin{bmatrix} 0 & 1 \\ 1 & 0 \end{bmatrix} \times \begin{bmatrix} 0 & 1 \\ 1 & 0 \end{bmatrix} = \begin{bmatrix} 1 & 0 \\ 0 & 1 \end{bmatrix}$$

You will see the result is a matrix which has ones in the top-left to bottom-right diagonal, and zeroes everywhere else. This is called the *identity matrix* and it is the matrix equivalent of "1" (because, if you multiply a matrix with the identity matrix then the original matrix is unchanged – just as if you were multiplying with 1). So the first of Dirac's requirements is satisfied (you might want to check that the square of matrix b is also the identity matrix).

The second of Dirac's requirements is that $(a \times b + b \times a)$ must equal zero. Let us test that this second requirement is satisfied:

_____

[3] Do you recognise these two matrices? They are actually two of the Pauli spin matrices we considered in the previous chapter. This is a big clue as to what is coming later.

$$a \times b = \begin{bmatrix} 0 & 1 \\ 1 & 0 \end{bmatrix} \times \begin{bmatrix} 1 & 0 \\ 0 & -1 \end{bmatrix} = \begin{bmatrix} 0 & -1 \\ 1 & 0 \end{bmatrix}$$

$$b \times a = \begin{bmatrix} 1 & 0 \\ 0 & -1 \end{bmatrix} \times \begin{bmatrix} 0 & 1 \\ 1 & 0 \end{bmatrix} = \begin{bmatrix} 0 & 1 \\ -1 & 0 \end{bmatrix}$$

So therefore:

$$(a \times b) + (b \times a) = \begin{bmatrix} 0 & -1 \\ 1 & 0 \end{bmatrix} + \begin{bmatrix} 0 & 1 \\ -1 & 0 \end{bmatrix} = \begin{bmatrix} 0 & 0 \\ 0 & 0 \end{bmatrix}$$

(Matrix addition is performed simply by adding the numbers which have the same positions in the two matrices).

You will see that the result of $(a \times b + b \times a)$ is the zero matrix (a matrix with zeroes in all its positions). So the second of Dirac's requirements is also satisfied by these two matrices.

So have we solved the problem? Have we found the relativistic equation of the electron?

Well, no, there is a problem.

According to relativity, space and time are placed on an equal footing and are combined to create *spacetime*. Spacetime is therefore four-dimensional (three dimensions of space, and one of time). But the analysis so far has only considered two dimensions: one dimension of space (a single value for momentum) and the time dimension. Remember, Dirac was trying to create a relativistic theory. So to be fully consistent with relativity, we have to consider the additional two space directions.

In the analysis so far, we had to find two matrices a and b such that either matrix squared equalled one, and also satisfied the requirement that $(a \times b + b \times a)$ must equal zero. Now we are moving to consider the two additional dimensions, we need to find a total of **four** matrices – a, b, c, and d – which satisfy the following requirements:

- a^2 must be equal to one, but b^2 (along with c^2 and d^2) must be equal to minus one.[4]

- If you take **any** two of the four matrices (let's call the two selected matrices x and y), then $(x \times y + y \times x)$ must equal zero.

(You will see that these requirements are virtually identical to the earlier situation in which we only considered two dimensions and two matrices).

However, it is simply not possible to find four 2×2 matrices which satisfy these requirements. Dirac found he could only satisfy these requirements by using larger 4×4 matrices. These important matrices are called the *gamma matrices*:

---

[4] Why should the sign be positive for a, but negative for b, c, and d? This is because distances in space and time do not work in quite the same way. The further you travel in space, the shorter the distance you travel in time. Hence, a space-travelling astronaut will age more slowly than his counterpart left on Earth.

The Gamma Matrices

$$a = \begin{bmatrix} 1 & 0 & 0 & 0 \\ 0 & 1 & 0 & 0 \\ 0 & 0 & -1 & 0 \\ 0 & 0 & 0 & -1 \end{bmatrix} \quad b = \begin{bmatrix} 0 & 0 & 0 & 1 \\ 0 & 0 & 1 & 0 \\ 0 & -1 & 0 & 0 \\ -1 & 0 & 0 & 0 \end{bmatrix}$$

$$c = \begin{bmatrix} 0 & 0 & 0 & -i \\ 0 & 0 & i & 0 \\ 0 & i & 0 & 0 \\ -i & 0 & 0 & 0 \end{bmatrix} \quad d = \begin{bmatrix} 0 & 0 & 1 & 0 \\ 0 & 0 & 0 & -1 \\ -1 & 0 & 0 & 0 \\ 0 & 1 & 0 & 0 \end{bmatrix}$$

This might be a good time to take your pen and paper out and check that these gamma matrices are correct and satisfy Dirac's requirements. Remember, a^2 (matrix a multiplied by itself) must be equal to one (the identity matrix). However, b^2 (along with c^2 and d^2) must be equal to minus one (the negative of the identity matrix).

Also, check that if you select **any** two of these gamma matrices (let's call the two selected matrices x and y), then $(x \times y + y \times x)$ must equal zero.

(Remember that i is the square root of -1. So if during the calculation you find you get a term which is i multiplied by i then the result is -1).

So, at last, these gamma matrices satisfy Dirac's requirements. And this is so important because – as we shall see – it is these gamma matrices which form the basis of the Dirac equation, and hence they describe the remarkable behaviour of electrons.

The equation in the Abbey

Once he had found his gamma matrices, Dirac then had to apply the final step in the creation of his relativistic equation of the electron: he had to "quantize" the equation, to make the equation agree with quantum mechanics. Fortunately, we already know how to do this as we studied it in the last chapter: it is simply to use the standard eigenvalue/eigenvector form which – as stated earlier – is used to describe all quantum mechanical measurements.

A concise form of the Dirac equation is shown below. You should recognise the standard form of the eigenvalue/eigenvector equation as described in the previous chapter. On the left hand side of the equation is an operator and a eigenvector (the state vector, ψ, considered in the previous chapter), and on the right hand side of the equation is an eigenvalue multiplying the same eigenvector:

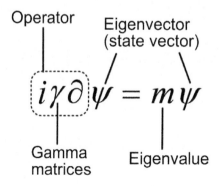

You will see that the set of the all-important gamma matrices (denoted by the Greek letter "gamma", γ) is included in the operator.

Paul Dirac died in 1984, and a commemorative plaque was installed on the floor of Westminster Abbey in London. You can see that the Dirac equation (which we have just derived) is inscribed on the plaque just below Dirac's name:

So, well done, we have derived the Dirac equation, one of the most important equations in physics. According to Nobel Prize winning physicist Frank Wilczek: "Of all the equations of physics, perhaps the most magical is the Dirac equation. It is the most freely invented, the least conditioned by experiment, the one with the strangest and most startling consequences. It became the fulcrum on which fundamental physics pivoted."

The only question left now is ... what does it all mean?

The absolute wonder

So now we have derived the Dirac equation, the relativistic equation for the electron, we have to determine what are the implications of the equation for the behaviour of the electron. We will find that the many implications are really quite remarkable, and explain the properties of the matter of which the world is made.

Firstly, if we examine three of the gamma matrices, we find that the 2×2 matrices in the top right segment of the gamma matrices precisely match the Pauli spin matrices (describing the spin of the electron) which we considered in the previous chapter:

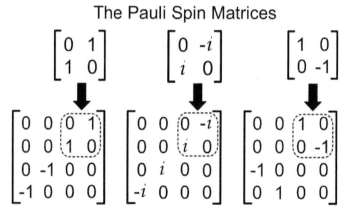

It appears that the gamma matrices seem to be describing the spin of the electron. Indeed, according to Dirac: "The gamma matrices are new dynamical variables which it is necessary to introduce in order to satisfy the conditions of the problem. They may be regarded as describing some internal motions of the electron, which for most purposes may be taken as the spin of the electron postulated in previous theories."

So this is really something quite remarkable. Dirac derived the need for electron spin **purely mathematically!** It is as if mathematics alone is predicting that the electron must be spinning (this derivation is in contrast to how we derived the Pauli spin matrices merely by **observing** electron spin and trying to model it mathematically).

This was truly an amazing achievement by Dirac. The physicist John Van Vleck compared Dirac's explanation of electron spin to "a magician's extraction of rabbits from a silk hat." According to Leon Rosenfeld, "the equation was immediately seen as **the** solution. It was regarded really as an absolute wonder."

This was perhaps the first time that mathematics on its own was used to predict the behaviour of Nature. It revealed the extraordinary power of mathematics as a tool for theoretical physics.

The spinning spinor

So we have now seen that the Dirac equation can be used to give us the Pauli spin matrices which describe the spin of an electron. And we have also seen that the Pauli spin matrices show that we will only measure an electron as being in one of two possible states: either "spin up" or "spin down".

Now let us consider what happens when an electron is rotated. We will find that the answer is quite astonishing.

In order to explain this behaviour, remember that in the previous chapter it was explained how the "spin up" and "spin down" states of the electron can be expressed as vectors. The "spin up" state can be described by the vector [1, 0] and the "spin down" state can be described by the vector [0, 1]. Let us draw those two vectors on a graph:

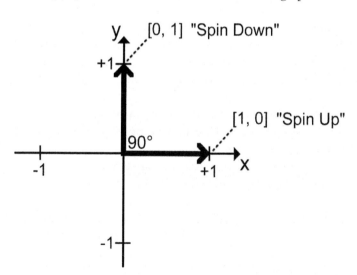

In the previous graph, you can see that the "spin up" state is denoted by a thick horizontal arrow from the graph centre to the coordinate [1, 0] ($x=1$, $y=0$), thus forming the "spin up" vector. You can also see that the "spin down" state is denoted by a thick vertical arrow from the graph centre to the coordinate [0, 1] ($x=0$, $y=1$), thus forming the "spin down" vector.

Note that the two vectors are at 90° to each other (for this reason, the two states are said to be *orthogonal*).

But here is the crucial point: in order to convert a "spin up" electron into a "spin down" electron we have to physically rotate the electron in real space by 180° (turning something pointing up into something pointing down). Amazingly, this suggests that if we physically rotate an electron by 180°, its quantum state (described mathematically by vectors) only gets rotated by 90°.

This reveals that the quantum state of an electron gets rotated by only half the amount that the electron is physically rotated. It is as if the physical electron is connected to its quantum state by a 2:1 gearing system. Bizarrely, this reveals that to return the quantum state to its original position (requiring a full spin of the quantum state by 360°), we have to physically rotate the electron by two full spins: 720°.

This is an amazing and crucial result: **an electron must be physically rotated twice, 720°, in order to rotate the quantum state by 360°, thus bringing the electron back to its original quantum state**.

This is really quite an extraordinary claim. Nothing in our everyday experience behaves like this: you don't have to rotate any ordinary object twice in order to bring it back to its original position – just one full turn is always enough! However, in the strange world of fundamental particles, we have to forget about our everyday experience and have an open mind about new modes of behaviour.

Now let us see what happens when we physically rotate our electron by 360°. Intuitively, we would expect a "spin up" electron to return to its initial "spin up" state when it is rotated by one full turn in physical space. However, we have just shown that a quantum state is rotated by only half the amount of the rotation of the physical electron. This means that the quantum state of the "spin up" electron is only rotated by 180°:

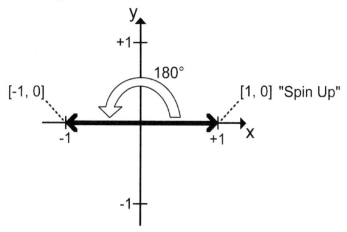

In the diagram above, you will see that the [1, 0] "spin up" electron vector (pointing to the right) has been rotated around the centre of the graph by 180° so it ends up pointing to the left. This represents the rotation of the physical electron by one full turn: 360°. But we can see that the coordinates of the resultant vector (pointing to the left) are [-1, 0]. We can see that these are not the coordinates of the "spin up" electron. In fact, they are the coordinates of the "spin up" electron **multiplied by -1** (because [1, 0] multiplied by -1 gives [-1, 0]). Most obviously, the vector is multiplied by -1 because it is the same length as the initial vector but **points in precisely the opposite direction.** It is the negative of the initial vector.

So, remarkably, this reveals that when we rotate an electron by a full 360° we do not end up with the same quantum state we started with. Instead we get the original quantum state multiplied by -1. **We will see in the next section that this is a hugely important result**.

This is an example of a mathematical object called a *spinor* (pronounced "spin-or" not "spine-or"). A spinor is an object which you have to rotate twice in order to return it to its original state. Needless to say, there is nothing in our everyday experience which behaves like this – we are obviously used to objects returning to their original state when we turn then by one full rotation. However, if you rotate a spinor by one full revolution, 360°, it is equivalent to multiplying the spinor by -1. If you then rotate it a further 360° you return the object to its original state (because -1 multiplied by -1 would give +1). So to return a spinor to its original state you have to rotate it by two full spins: 720°.

A particle (such as an electron) which you have to rotate twice to return to its original state is called a *spin-½* particle (you might say a full rotation only gets you halfway!).

Rather wonderfully, this is best illustrated by a practical example. It so happens that the rotation of the elbow and shoulder in the human arm is a perfect model of a spinor (there are some undoubted evolutionary advantages to this spinor design, allowing the hand to rotate fully without the arm getting tangled). I have recorded a video showing me using my arm to rotate a glass of water which represents an electron.

I can recommend you watch the video at http://tinyurl.com/particlerotation

On the opposite page are some still images from the video:

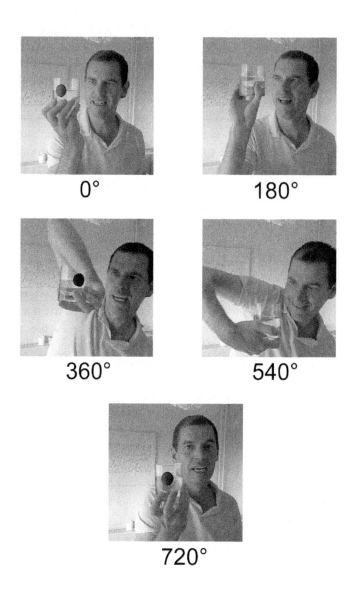

Though it might sound almost unbelievable that spin-½ particles have to be rotated by 720° to return to their original state, it has been directly experimentally proved. A stream of neutrons, whose spin was all oriented in the same direction, was split into two beams. One stream was rotated about an axis along its direction of motion, and then the two beams were recombined. When the rotation angle was 360°, the two beams cancelled-out. Amazingly, though, when the particles were rotated twice, through 720°, the two beams combined to produce a stronger signal. This shows the peculiar rotational behaviour of spin-½ particles is a genuine measurable effect.

The Pauli exclusion principle

All spin-½ particles are called *fermions*. Fermions are the particles which form atoms, and hence form all the matter in the universe. But why should spin-½ particles be the matter particles? Well, we will now discover that the reason that fermions behave like matter particles is directly linked to their peculiar rotation characteristics.

The secret lies in what happens when we exchange two particles. Imagine we have two fermions of the same type: particle A and particle B (see the following diagram). These might be electrons, for example. We want to exchange (swap) the two particles. The most obvious way we could do this would be by moving particle A to the position of particle B while, at the same time, moving particle B to the position of particle A:

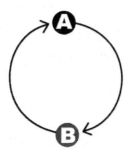

However, this is not the only way we could exchange the particles. It is possible to exchange the particles using a different method while still producing exactly the same result. As shown in the following diagram, we could translate both particles vertically, and then **rotate the whole system around the centre of particle B**. This would produce exactly the same end result, with particle B now above particle A:

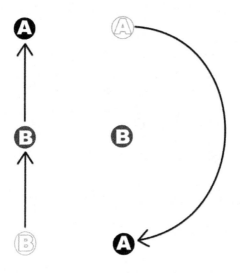

But what happens to the quantum state (the wavefunction) of the system when the system is translated and rotated like this? Well, firstly, the translation has no interesting effect. So that means we have done something extremely clever here: we have converted the process of particle exchange into a process of simple rotation of particles: we have shown that the mathematics of particle exchange and particle rotation is essentially the same. And we know precisely how fermions behave when they are rotated as we worked it out in the previous section. We have seen that fermions behave quite peculiarly when they are rotated. So we will have to examine the impact of the rotation on the quantum state of the system.

It can be seen that particle A was rotated by 180°, and particle B was also rotated by 180° (around its centre). The total rotation as far as the quantum state of the system is concerned is the sum of the rotations of the two particles, which is therefore 360° (for more details of this approach to particle exchange and rotation, see the footnote[5]).

This is a crucial result. Because, as saw in the previous section, a 360° rotation of a spin-½ particle results in the quantum state being multiplied by -1. It is said that the wavefunction in that case is *antisymmetric* (because it does not join up with itself when it is rotated: it is multiplied by -1). We can now see that particle exchange is also antisymmetric: if you exchange two fermions, the quantum state is multiplied by -1.

---

[5] For more details of this approach, see Chapter Twelve of *Quantum Field Theory and the Standard Model* by Matthew D. Schwartz.

(If you continue and exchange the particles back again, the quantum state is multiplied by -1 again, thus restoring the original quantum state because -1 multiplied by -1 gives +1).

Now let us consider the situation when two fermions which are in the same state are exchanged. Firstly, we will have to consider what does it mean for two particles to be in the same state.

As described in Chapter One, all particles of a particular type are identical: every electron is the same as every other electron. But does this mean that all electrons are indistinguishable from each other? No, of course not. There are obviously many separate electrons in the universe, and we (and Nature) can distinguish each electron from every other electron because they are in different positions, or they might be moving at different velocities. So even though every electron is **identical**, electrons are not **indistinguishable**.

There are only a few ways in which particles can be distinguished (e.g., their position, their momentum) and these distinguishing features are called *degrees of freedom* (considered in my third book).

For electrons orbiting the same atomic nucleus, there are only a very few ways in which those electrons can be distinguished from each other (they have very few degrees of freedom). As a result, the possibility arises that two electrons might be in exactly the same state. If those two electrons are exchanged, there clearly cannot be any overall change to the quantum state of the system because the electrons are **completely indistinguishable**. Nature has fundamentally no way of distinguishing them. If the electrons are exchanged, nothing can possibly change.

However, as we have just seen, if any two fermions are exchanged then the quantum state of the system must be multiplied by -1. These two results – one saying nothing can possibly change, one saying the quantum state must change – are clearly contradictory. In fact, the only number which does not change when it is multiplied by -1 is the number zero. It would appear that the system can only be described by a wavefunction with the value of zero. We can come to only one conclusion: this contradictory state cannot exist.

We have derived the *Pauli exclusion principle*: **two identical fermions cannot occupy the same quantum state**.

As Jim Baggott says in his book *Higgs*:

> *The principle derives from the mathematical form of the wavefunction for any composite state consisting of two or more electrons. If the composite state were assumed to be created with two electrons which have precisely the same physical characteristics, then the wavefunction has zero amplitude – such a state could not exist.*

Atoms are mainly empty space, so why can't I push my finger through a wooden desk? Isn't there enough empty space for two atoms to pass through each other? Well, it is the Pauli exclusion principle which brings solidity to matter. I cannot push my finger through a wooden desk because any penetration of my finger would result in the electrons in the atoms in my finger being in the same state (same position) as the electrons in the atoms in the wood. The exclusion principle states that is simply not possible. Hence, the wood appears impenetrable to your finger.

So the Pauli exclusion principle results in fermions taking the role of the matter particles, the particles which compose all of the objects in the material world. Cars, trees, water, gases, human beings, are all composed of fermions: protons, neutrons, and electrons.

The root of all chemistry

So, we have now seen that no two fermions can occupy the same state. This is definitely going to have an impact on electrons in a atom orbiting the atomic nucleus.

If you remember back to the discussion in Chapter One, electrons can orbit the nucleus in only certain quantized orbits (or *shells*), the shells being nested around each other like Russian *matryoshka* dolls. The shell in which an electron resides is determined by the energy of that electron. It would appear that any two electrons in the same shell would be in the same state (possessing the same energy). Therefore, according to the Pauli exclusion principle, this would appear to prohibit each shell from containing more than one electron.

However, there are a few other ways in which the state of orbiting electrons can be distinguished. This is because the spherical shells are not necessarily completely identical. Essentially, they can wobble in an elliptical manner (think of a soap bubble wobbling as it rises in the air). These deformations of the shells provide us with a few more ways to distinguish the states of electrons which possess the same energy (i.e., the electrons have a few additional degrees of freedom).

In fact, an electron orbiting the nucleus can be defined by three quantum numbers: n (the shell number, determined by the energy of the electron), l (the shape of the shell), and m (the direction in which the shell is pointing). There are strict rules as to what values these numbers can take. For a given shell number, n, the rules are:

- l can take any integer value from 0 to (n-1).

- m can take any integer value from -l to l.

You will see that these two simple rules can be used to construct the following table (which considers the first three shells in which n is 1, 2, or 3). What we are interested in is how many unique combinations of the three numbers there can be, because that tells us how many electrons (each being in a unique state) there can possibly be in each shell. You can clearly see from the table just how many unique combinations there can be for any value of n:

n	l	m	Number of combinations for a particular value of n
1	0	0	**1**
2	0	0	
2	1	-1	
2	1	0	**4**
2	1	1	
3	0	0	
3	1	-1	
3	1	0	
3	1	1	
3	2	-2	**9**
3	2	-1	
3	2	0	
3	2	1	
3	2	2	

This table suggests that there is only one unique combination of the three quantum numbers in the first shell, there are four combinations in the second shell, and nine unique combinations in the third shell. In other words, the number of combinations is given by n^2.

This number should also represent the number of electrons it is possible for a shell to hold (because each of those electrons would be in a unique state). In other words, the lowest energy shell should be capable of holding just one electron, the second shell should be capable of holding four electrons, and the third shell should be capable of holding nine electrons.

However, this is not quite right as there is one additional degree of freedom possessed by each electron, and that is the value of electron spin. Remember, each electron also has the freedom to be either "spin up" or "spin down". These two options means that the number of electrons it is possible for each shell to hold must be multiplied by two. The formula therefore becomes $2n^2$.

This doubling means we should now expect the first shell to be capable of holding two electrons, the second shell should be capable of holding eight electrons, and the third shell should be capable of holding eighteen electrons. And so on.

This, finally, is correct. The number of electrons each shell can hold is 2, 8, 18, 32, 50, ...

Let us see what this means in practice by considering the structure of some chemical elements.

Hydrogen is the lightest element, having an atomic number of 1. This means that an atom of hydrogen has just one proton – and one orbiting electron. However, oxygen has an atomic number of eight, which means it has eight orbiting electrons. All of these eight electrons cannot fit into the lowest energy shell because we know only a maximum of two electrons can fit into that shell. Therefore, two electrons go into the lowest shell, and the remaining six electrons have to go into the next higher shell:

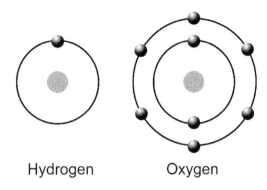

Hydrogen   Oxygen

The important factor is now the number of electrons in the outer shell of these atoms. These are called the *valence* electrons. We can see that the hydrogen atom has one valence electron, while the oxygen atom has six valence electrons. This raises the possibility of two hydrogen atoms sharing their valence electrons with one valence electron of the oxygen atom:

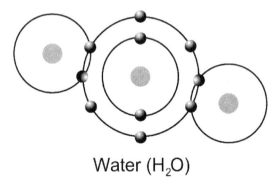

Water (H_2O)

In the diagram above we can see that each of the two hydrogen atoms is sharing its valence electron with one of the valence electrons of the oxygen atom. In this way, each hydrogen atom effectively gains one more electron in its outer shell (thus filling its outer shell, which can only contain two electrons), and the oxygen atom effectively gains two electrons in its outer shell (thus filling its outer shell, which can contain eight electrons).

This forms a molecule of water: H_2O (two atoms of hydrogen, one atom of oxygen).

This method of sharing valence electrons is called *covalent bonding* and is the structural method which forms molecular compounds such as carbon dioxide, methane, glucose, or even buckminsterfullerene:

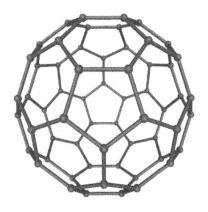

Buckminsterfullerene's "bucky ball" structure is an example of an *allotrope* of carbon, just one example of how sixty carbon atoms can join together to form a molecule (which resembles a microscopic football). Other allotropes of carbon (different structural arrangements) are diamond and graphite, and also *graphene* – hexagons of carbon atoms which join together to form a layer which is just a single atom thick (an illustration of graphene is shown on the front cover of this book). Graphene is acclaimed as the new wonder material, 200 times stronger than steel, yet transparent and flexible.

However, some atoms have full outer shells of electrons. These are the so-called *noble gases*: helium, neon, argon, krypton, xenon, and radon. Because they have full outer shells, atoms of these gases do not combine with other atoms to form molecules. This makes them chemically inert.

We can see that the Pauli exclusion principle – and the numbering of electrons in the shells of atoms – is therefore responsible for the whole of chemistry. As Jim Baggott says in his book *The Quantum Story*:

> *By preventing the electrons from collapsing or condensing into the lowest-energy orbit, the exclusion principle allows complex multi-electron atoms to exist in the pattern described by the periodic table. It enables the existence of a marvellous variety of elements, the multitude of possible chemical combinations, and hence all material substance, living and non-living. This was a fantastic achievement.*

We have covered a great deal of ground in this chapter, and some of the material was challenging, but I hope you can now see that it was worth it. We have derived the Dirac equation, shown how it predicts the spin of electrons, shown how the antisymmetric nature of that spin leads to the Pauli exclusion principle, and finally shown how the Pauli exclusion principle leads to the solidity of matter and prevents all the electrons in an atom collapsing to the lowest energy shell – thus giving atoms their structure. It is that structure which leads to the whole of chemistry. The Dirac equation undoubtedly deserves a book all to itself.

And we have not even considered the most extraordinary prediction of the Dirac equation …

4

ANTIMATTER

The date is 30th June 1908.

Sergei Semenov had just finished his breakfast and was sitting outside his house, enjoying the cool air before the summer sun rose in the sky to bring the extreme summer heat. Occasionally, hunters and fishermen would pass by, but usually his only neighbours in this remote region of Siberia were the bears and deer who roamed the endless pine forest.

As Sergei peered up into the cloudless sky that morning, suddenly he saw the sky split in two, divided by a towering wall of flame. The heat was so intense that Sergei felt as though his shirt was on fire. However, before he could rip his shirt off, a shockwave hit Sergei, blowing him off his feet and sending him sprawling several metres away.

Sergei's wife ran out of the house, in a state of panic. When she saw her husband virtually unconscious, she dragged him back into their house. But there was no safety to be found there. The earth shook, and rocks started falling from the sky like fiery cannon balls. Sergei and his wife cowered on the floor of their home, waiting for what was surely the end of the world.

Thirteen years passed. World War One, and the 1917 Russian Revolution, all passed without impacting on remote Siberia in any way. Then, in 1921, a mineralogist Leonid Kulik based in St. Petersburg heard the rumours of the catastrophic event, apparently centred on the Tunguska river valley, in one of the most remote regions of Siberia. Kulik believed the event was caused by a meteorite impact, and managed to persuade the Soviet Academy of Sciences to pay for a mission to recover iron from the meteorite which could be used by Soviet industry.

Kulik's team faced an arduous journey to the site of the explosion, carrying heavy equipment largely by hand as they fought through uncharted forest and swamp. When Kulik arrived at the site he was amazed to find scorched trees flattened like matchsticks over an area about the size of Greater London, but he found no impact crater (which would be expected for a meteorite of that size).

Here is one of Kulik's original photographs:

Kulik drained many of the swamps in the area, but found no evidence that any of them were impact craters. He plotted the position of every tree, but no evidence of meteorite fragments were found. Kulik continued to investigate Tunguska for a further 20 years without finding evidence of a meteorite impact. When the Nazis invaded Russia in 1941, Kulik volunteered to fight for the Red Army. He was captured by the Germans, and died a year later.

Asteroid *2794 Kulik* was named in honour of Leonid Kulik.

Meteorite impact remains the most likely explanation for the explosion. However, Prof. Frank Close of Oxford University has another explanation. In his book *Antimatter* he considered the theory that a block of antimatter might have been the cause of the explosion. Antimatter is perhaps best known as the power source behind a *Star Trek* warp drive, but it is most certainly a real substance possessing incredible power. Antimatter is the precise opposite of everyday matter which, for example, makes cars and trees. Everyday matter is made of fermions (*fermionic* matter), the particles we have already considered. Antimatter is made of *antifermions*.

When antimatter comes into contact with matter, they annihilate each other in a blinding flash of energy (gamma rays, high-energy light). Antimatter releases more explosive energy than anything else in existence. An antimatter bomb would be 1,000 times more efficient (in terms of fuel) than the fission bombs dropped on Hiroshima and Nagasaki. An antimatter power source would be 100 times more efficient than nuclear fusion.

Undoubtedly, antimatter is being taken seriously by the military as the ultimate weapon of mass destruction. In 2004, Kenneth Edwards, the director of the Munitions Directorate at Eglin Air Force Base in Florida, gave a keynote speech at the NASA Institute for Advanced Concepts. His speech stressed that even small granules of antimatter – almost invisible – could possess devastating destructive potential. Just 50 millionths of a gram of antimatter would generate a blast equal to the explosion at the Federal Building in Oklahoma City in 1995, which killed 168 people.

The theory that a lump of "antirock", one metre across, was the cause of the Tunguska blast is based on the fact that no solid evidence of a meteorite or its crater has ever been found. Whatever caused the Tunguska blast disappeared into thin air – just like an antimatter/matter annihilation. The antirock theory might just be one of many highly speculative suggestions about the source of the Tunguska explosion, suggestions which include the idea that a missing 1972 atomic bomb might have dropped through a wormhole and travelled backward in time to 1908. However, at the very least, the antimatter theory draws our attention to the undoubted vast power and potential of the mysterious substance known as antimatter.

Angels and demons

Antimatter also plays a central role in Dan Brown's bestselling book (and movie starring Tom Hanks) *Angels and Demons*. Though the plot of *Angels and Demons* is fairly preposterous, the science related to antimatter is accurate. Indeed, CERN (the largest particle physics laboratory in the world) has even created its own *Angels and Demons* website:

http://angelsanddemons.web.cern.ch

You will notice the Dirac equation is featured at the top of the following page on the CERN webpage. This is a clue as to the nature of antimatter:

http://angelsanddemons.web.cern.ch/antimatter

Warning: spoilers ahead. The plot of *Angels and Demons* considers a rogue scientist who secretly uses the LHC (Large Hadron Collider, based in CERN) to create a quarter of a gram of antimatter. The antimatter was suspended in a vacuum by a magnetic field to stop it annihilating and exploding by touching the sides of the canister (this is all good science: CERN has trapped antimatter for 57 days using a similar method – a world record).

In the book, the antimatter is used to form a bomb which threatens the Vatican City. The quarter gram of antimatter is equated to an explosive power of five kilotons of TNT (about a quarter of the power of the Hiroshima nuclear explosion).

The main character in the book, Robert Langdon (played in the movie by Tom Hanks), then has to pursue many clues in a frantic chase to discover the antimatter before it destroys the Vatican. In this respect, the task of Robert Langdon is similar to the task of a physicist: solve a series of puzzles and follow a series of clues to get to the bottom of things.

As an example, some of the clues which Langdon has to unravel take the form of *ambigrams*, which are symmetric words which read the same when they are turned upside down. An example of an ambigram version of the word "Antimatter" is:[6]

Yes, turn the book upside down and you will see that the word is still the same upside down. How remarkable.

So when we do physics we have to think like Robert Langdon. We have to look for clues, and solve mysteries.

---

[6] Custom ambigram designed specially for this book by FlipScript Corporation.

As an example, let us consider one of Paul Dirac's gamma matrices. We might stare at it for a while, and wonder about its significance. We might even turn our book upside down in desperation. In which case we would notice something highly significant …

… the gamma matrix is an ambigram!

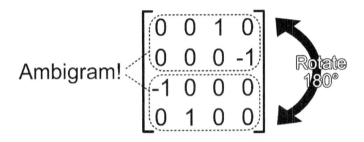

Just like one of Robert Langdon's symmetric words, if you rotate this book 180°, turning this gamma matrix upside down, then you find the gamma matrix looks the same. All the ones and zeroes appear in the same position. We have discovered a vital clue. This property of the gamma matrices surely has some deep underlying significance.

We know that one half of a gamma matrix gives us one Pauli spin matrix, and we know that that Pauli spin matrix describes an electron which is either "spin up" or "spin down". But what does the other half of the gamma matrix represent?

When the Dirac equation was derived in 1928, it was quickly realised that the "other half" of the gamma matrices had to represent negative energy states of the electron. If we consider the energy-momentum equation again (which Dirac used to derive his equation) we see it is a formula for the square of the energy, E^2. That meant that the actual value of E could be a positive value **or a negative value**, because the square of a negative value also gives a positive result. However, no negative energy had ever been observed. A stationary object has zero energy of motion, and once it starts moving its energy only increases. To talk of negative energy seemed absurd.

Many physicists criticised the Dirac equation on the basis of its prediction of negative energy states. In response, Dirac felt obliged to defend his theory, and presented a possible solution which made use of the Pauli exclusion principle. Dirac suggested that all of the negative energy states had already been filled by a vast "sea" of electrons. This is not unreasonable, as electrons would naturally seek the lowest energy state, and a negative energy state is, of course, lower than any positive energy state. The Pauli exclusion principle then states that no two electrons can be in precisely the same state, so with all the negative states occupied there would be no more room for other electrons to enter negative states. The remaining electrons would effectively be forbidden from having negative energy.

Dirac went even further. He suggested that any unoccupied "hole" in the Dirac sea of negative energy states would effectively represent a positive charge (because it would require a negatively-charged electron to fill the hole). Hence, Dirac suggested that these "holes" might be positively-charged protons. It was only when it was realised that a proton is approximately 2,000 times heavier than an electron that Dirac shelved that idea.

By 1931, most physicists had lost interest in the "Dirac sea" proposal, and even Dirac appeared to have lost confidence in his "hole" theory as well. Instead, Dirac now predicted something much more solid:

A hole, if there was one, would be a new kind of particle, unknown to experimental physics, having the same mass and opposite charge to an electron. We may call such a particle an anti-electron.

Paul Dirac had just predicted the existence of antimatter.

The discovery of antimatter

Before the days of particle accelerators, experimental particle physicists used a *cloud chamber* for discovering new particles. A cloud chamber is a sealed glass container containing water vapour. Any charged particle which passes through the cloud chamber leaves a thin trail of condensed water droplets in its wake, like vapour trails behind a jet aircraft. By measuring the curvature of the trail under the influence of a magnet, it is possible to tell if the particle is positively or negatively charged (the particles curve in different directions). The paths of lighter particles are curved by a greater amount.

The Earth is constantly bombarded by *cosmic rays*, which are high-energy particles from outer space. These are actually an excellent source of particles for experimental purposes, with some of these particles possessing far greater energies than can be produced by our largest particle accelerators. Cosmic rays can be detected by cloud chambers, leaving wispy ghost-like trails as they pass through.

In 1932, a research student at Caltech called Carl Anderson was using a cloud chamber when he noted something unusual. Electrons are light particles, and leave thin wispy trails through a cloud chamber. Anderson saw several particles leaving these wispy trails, but they curved in opposite directions. This appeared to indicate that some of the electrons had positive charge. This is precisely what would be expected of the mirror-image antimatter version of the electron: the *positron*.

Antimatter had been detected. Dirac's theory had been confirmed. Carl Anderson received the Nobel Prize in 1936 for his discovery.

Dirac's equation applies to all fermions, not just electrons. So a proton also has an antiparticle, called the *antiproton*. And the antiparticle of the neutron is the *antineutron*. In fact, we now know that every particle has an antiparticle.

Antimatter can be produced naturally by radioactive sources. Even a banana – which contains potassium-40 – produces one positron about once every 75 minutes. Antimatter can also be produced in particle accelerators, though the amounts are extremely small: the total produced by CERN is only about 15 nanograms. If all the antimatter produced by humans was annihilated at the same time, the energy produced would not be enough to make a cup of tea.

Because it is produced in such small quantities, antimatter is the most expensive substance in existence. Gold is worth 56 dollars per gram. Platinum is worth 60 dollars per gram. Rhino horn is worth 110 dollars per gram. Heroin is worth 130 dollars per gram. All of these substances are expensive, but antimatter is slightly more costly, at 6.25 trillion dollars per gram.

Antimatter has found a use in medical science. In Positron Emission Tomography (PET) scanning, the patient is injected with a radioactive positron source. When the positrons are emitted, they are immediately annihilated by the surrounding matter (in the tissue). This annihilation produces gamma rays (high-energy light). The gamma rays are actually composed of two photons, travelling in precisely opposite directions (due to conservation of momentum). If the patient is placed inside a circular detector, a line can be drawn between the two detected photons. This allows very accurate three-dimensional models of the tissues of the body to be constructed on a computer.

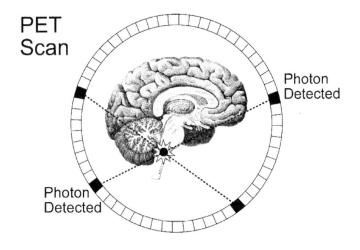

Backward in time

Richard Feynman is regarded as one of the greatest physicists of the 20th century. He is also regarded as one of the most flamboyant, with a love of playing the bongo drums, and a gift for explaining complex ideas in an imaginative and entertaining manner. Feynman worked on the Manhattan Project, developing the atomic bomb, but it was for his work on developing a quantum explanation of the electromagnetic force that he obtained his Nobel Prize in 1965 (we will be considering this in detail in Chapter Six).

As part of Feynman's solution, he developed a series of simple diagrams to represent interactions between particles. These are called *Feynman diagrams*. As an example, all electromagnetic phenomena can be reduced to the simple interaction shown in the following Feynman diagram:

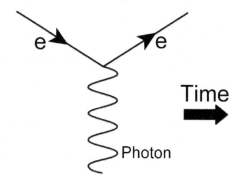

The diagram represents a single *vertex* from a Feynman diagram. In the diagram, you will see that there is a time axis flowing from left to right. The vertical axis represents space. So this particular Feynman diagram represents an electron coming in from the left side of the diagram, and emitting a photon (denoted by the wavy line), with the result that the path of the electron is altered. This is due to conservation of momentum: imagine you are ice skating, and you throw a heavy object to one side, your path would be modified to the opposite direction.

Back to the diagram, the theory of relativity tells us that we can treat time as just another dimension – just like the three dimensions of space. So, rather wonderfully, this gives us the freedom to rotate a Feynman diagram to any orientation (effectively exchanging time and space) and the interactions which result will still be valid.

As an example, the following diagram shows the previous Feynman diagram rotated 90° anticlockwise:

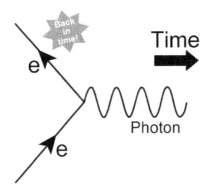

This diagram now reveals something remarkable. Let us analyze it.

We still see the bottom electron entering the diagram from the left, and it is moving in the forward time direction. But the top electron now appears to be moving in the reverse time direction! Surely that cannot be correct, can it?

Well, let us imagine the top electron is actually an anti-electron: a positron. In that case, we can re-interpret the diagram as showing an electron and a positron coming in from the left, and annihilating each other releasing energy (the photon). So the diagram makes sense if we interpret the top electron as a positron.

But the top electron is clearly moving in the backward time direction. Remarkably, what this reveals is that we can interpret a positron (an anti-electron) as an electron moving in the backward time direction!

It is quite an amazing result, but it is now generally accepted: a positron is an electron which is moving backward in time. (This ties in nicely with the theme of my third book, which explored the idea of negative energy as motion in the backward time direction).

The origin of quantum field theory

Let us see what other remarkable things we can discover from this Feynman diagram. Let us continue to rotate our Feynman diagram, this time by a further 180°:

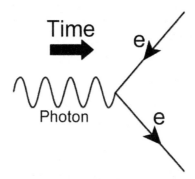

Again, this is quite remarkable. We now see a photon coming in from the left, and the photon producing a pair of particles: an electron and a positron (the top electron which is travelling backward in time). The photon is not a fermion, it is not a matter particle, so we can interpret this diagram as the production of a pair of matter particles from – essentially – pure energy. It is clear from this that the number of matter particles in the universe is not fixed.

What is even more remarkable is that – according to the uncertainty principle of quantum mechanics – it is possible for a small amount of energy to appear essentially "out of nothing" (as long as it is only present for a short time). You might think of this as a short-term bank loan: you can borrow money from nothing, as long as you pay it back after a short time. This means that energy (and, therefore, matter particles) can appear essentially "out of nothing".

This, then, changes the way we have to look at particles – and empty space. Instead of considering particles as simply wandering around in empty space, we now see that particles are being constantly produced and annihilated in space. Space is clearly not empty, but is a seething mass of particles appearing and disappearing. Instead of "empty" space, we now consider space as being the *vacuum*. The vacuum is far from empty.

In fact, if the number of particles is not fixed, and particles can easily appear and disappear, then this underlying vacuum or field of space should be considered as being "more fundamental" than the particles themselves. It is the properties of the vacuum that should be considered, as it is the constant presence of the vacuum which truly controls the particles. Essentially, this represents a move away from a particle-based theory toward a field-based theory.

As Roger Penrose says in his book *The Road to Reality*:

> The key property of an antiparticle is that the particle and antiparticle can come together and annihilate one another, their combined mass being converted into energy, in accordance with Einstein's $E=mc^2$. Conversely, if sufficient energy is introduced into a system then there arises the strong possibility that this energy might serve to create some particle together with its antiparticle. Thus, our relativistic theory certainly cannot just be a theory of single particles, nor of any fixed number of particles whatever. Indeed, according to a common viewpoint, the primary entities in such a theory are taken to be the quantum fields, the particles themselves arising merely as "field excitations".

This reveals a move away from a quantum theory of particles to *quantum field theory*. Later in this book we will examine how quantum field theory provides us with an explanation of the fundamental forces between particles.

This brings us to the end of the first part of this book. We have considered the structure of the atom, specifically "the forces which hold the atom apart". We have seen how energy quantization prevents electrons from spiralling into the nucleus. We have also seen how the Pauli exclusion principle prevents all the electrons from occupying the ground state – the lowest energy level – thus giving the electron shells of atoms their characteristic form.

We will now move onto the second part of the book, "the forces which pull the atom together". And we will start by considering two of the most important principles in physics.

5

SYMMETRY AND CONSERVATION

In this chapter we will be considering some of the most valuable tools in the particle physicist's toolkit: the laws of conservation, such as the law of conservation of energy. We will also be uncovering a remarkable and surprising link between conservation and symmetry, a link which has proven to be remarkably fruitful in the attempt to make sense of the behaviour of elementary particles.

As Paul Davies says in his introduction to Richard Feynman's book *The Character of Physical Law*: "A great unifying theme among particle physicists has been the role of symmetry and conservation law in bringing order to the subatomic zoo."

What is a law?

In human terms, we might have laws restricting where we can park our cars, or maybe laws prohibiting us from stealing from supermarkets. These laws have been developed for the good of all society, and most people abide by those laws, maybe because of a sense of moral duty, or maybe because of a fear of punishment for any transgression. However, if we so wished, we could choose to break any of these laws.

In this respect, the laws of Nature are different from human laws. Try breaking the law of conservation of energy! You will have great difficulty (hint: it's not possible).

And the laws of Nature do not just constrain the behaviour of humans – they constrain the behaviour of everything in existence. The behaviour of all atoms, all particles, must abide by these fundamentally unbreakable laws of Nature. As Vincent Icke describes in his book *The Force of Symmetry*:

> *The essential uncertainty of quantum behaviour might create the uneasy suspicion that anything goes in this world. But since the universe has a very definite structure, it seems very unlikely that everything is allowed.*
>
> *As it happens, there are many things that are forbidden, and it is the forbidding rules that give structure to the world.*

You may well be aware of conservation laws such as the law of conservation of energy, or the law of conservation of momentum. A conservation law says that some measurable property of a system is maintained at a constant value over a period of time. For example, consider two objects colliding with each other. If you calculate the total momentum of both objects before the collision, you will find the total value is the same after the collision. This is an example of the law of conservation of momentum.

The law of conservation of energy is very similar. If you calculate the total energy before the collision, you will find that total value is unchanged after the collision. Another way of stating this principle is to say that energy cannot be either created or destroyed.

However, you might very well discover that the total kinetic energy of the objects (the energy associated with the movement of the objects) is not maintained after a collision.

This is because energy can take different forms. There is energy associated with heat, and with sound, and with radiation. If the objects produce a tremendous crash when they collide, then some of the kinetic energy has been converted into sound, and probably into heat as well. The total kinetic energy might well be reduced, but if you could calculate the total of all forms of energy after the collision then you would, indeed, find that the total amount of energy had been conserved.

Brian Cox explained this conservation principle in his BBC TV Series *Wonders of Life*:

> Over the years, the nature of energy has proved notoriously difficult to pin down, not least because it has the seemingly magical property that it never runs out – it only ever changes from one form to another. The key thing is, energy is conserved: it is not created or destroyed.
>
> The fact that energy is neither created nor destroyed has a profound implication. It means energy is eternal. The energy that's here now has always been here, and the story of the evolution of the universe is just the story of the transformation of energy from one form to another. Every single joule of energy in the universe today was present at the Big Bang 13.7 billion years ago.

Several conservation laws have been discovered, most notably the law of conservation of energy, conservation of momentum, conservation of angular momentum, and conservation of electric charge. The first three of these conservation laws demand that a particular quantity (for example, the amount of energy) in a system remains constant over time. However, the fourth of these laws – the conservation of electric charge – is an example of a different type of conservation law in that it simply involves the "counting" of objects. Some elementary particles – for example, electrons – have negative electric charge. Some different elementary particles – for example, positrons – have positive electric charge. The charge conservation law requires a bit of counting: add up the number of particles with positive charge, subtract the number with negative charge, and you will find the total number does not change with time.

(The "counting" conservation laws are perhaps the most intuitive of the conservation laws. Place a number of balls in a bag. Come back three days later and you will still find the same number of balls in the bag. We take these things for granted, but that is a conservation law in action! Richard Feynman simply referred to a law of "conservation of objects", though we might consider it being fundamentally due to a law of conservation of mass and energy.)

Why are the conservation laws so extremely valuable to particle physicists? Well, if you know that some total property of a system of particles is going to be conserved, you can use this information to infer which type of particles can possibly be produced, for example, during a collision. If you find your sums do not add up – for example, if you have a surplus of energy left over after the collision – you might infer that an additional particle had been produced to account for the surplus energy.

But bear in mind that any additional particle would have to satisfy **all** the conservation laws. In other words, it would have to comply with the law of conservation of energy, and conservation of momentum, and of spin, and of charge. So all these conservation laws impose considerable constraints on particle production.

Let us examine this more closely by considering an example of *particle decay*. Particle decay occurs when a particle spontaneously (and randomly) transforms into other particles. As an example, the process of beta radioactive decay involves the decay of a neutron – in the nucleus of an atom – into a proton. An electron is also emitted in the process, and it is this electron which constitutes the beta radiation emitted by the nucleus.

However, when physicists applied the law of conservation of energy to this decay process, they found something was missing. The energy of the resultant proton and electron did not add up to the energy of the initial neutron. So where did the excess energy go?

In 1930, Wolfgang Pauli proposed that the excess energy went into creating an additional emitted particle. This would ensure that the energy of the particle going into the interaction matched the energy of the particles coming out of the interaction. This gives you an idea of how the conservation laws are a vital tool which can be used to predict the existence of new particles.

But what else did the conservation laws say about this new particle? Well, let us consider the law of conservation of electric charge. The neutron going into the interaction did not have any electric charge (hence its name). Coming out of the interaction, the proton had a positive charge, but this was balanced by the electron coming out of the interaction which had a negative electric charge. So the law of conservation of electric charge was already correctly balanced: no charge going in, no net charge coming out. This meant that the hypothetical new particle could have no electric charge or that would cause an imbalance. For this reason, in 1934 Enrico Fermi gave the new particle the name *neutrino* (a "small neutral thing" – in Italian).

The neutrino was eventually discovered in 1956, but its existence and properties had been predicted twenty years earlier thanks to the conservation laws.

Particle accelerators

In a particle accelerator, electrically-charged particles are accelerated by electric forces (the modern descendent of Ernest Rutherford's cathode ray tube). A magnetic field is then used to curve the path of those particles, and this can result in particles travelling around a large loop, accelerating to a speed close to the speed of light. The particles are then made to collide with a variety of targets. By examining the results, we can learn about the structure of matter.

The Large Hadron Collider (LHC) is the largest particle accelerator ever made. In fact, it is the largest machine ever made. The LHC is a 27-km circumference loop which straddles the French-Swiss border. It is buried 175 metres deep (not for any scientific reason – only because it was simply cheaper than buying land in Geneva).

The purpose of the LHC is to accelerate beams of protons to nearly the speed of light, and then to collide two beams and examine the results.

Nothing can move faster than the speed of light, so the speed of light forms the upper speed limit for accelerated particles. However, the energy of a particle increases sharply as it approaches (but does not exceed) the speed of light. An accelerated proton in the LHC can move at 99.9999991% of the speed of light, at which point its energy is 7 TeV ("eV" represents "electronvolt" – a small unit of energy which is the standard measure of energy in particle accelerators, with 1 TeV being equal to a trillion electronvolts). An energy of 1 TeV represents the energy of a single flying mosquito. So at the LHC, the energy of 7 flying mosquitos is packed into a single proton (about a million million times smaller than a mosquito). This represents the current world record for particle accelerator energy.

The two beams of protons travel around the LHC in opposite directions before colliding head-on, giving a total collision energy of 14 TeV (7 TeV per beam).

The 27-kilometre circumference of the LHC might sound a long way, but a proton accelerated to nearly the speed of light will make over 11,000 circuits of the loop each second. At this speed, the entire beam of protons has the same energy as a French TGV train travelling at top speed. In fact, the protons are travelling so fast that in their 10-hour lifetime in the accelerator they travel 10 billion kilometres: equivalent to the distance to Neptune – and back. In ten hours!

So what is the purpose of these extraordinary particle accelerators? Why must the collision energies be so large?

There is, perhaps, a general misconception as to the purpose of these machines. It is not helped by the use of the term "atom smasher" to describe them. The impression is often that these machines need high collision energies to "break" particles into "smaller and smaller pieces". However, as was described in Chapter One, elementary particles are believed to be infinitely small points, with no internal structure – no constituent parts. It is simply not possible to break them down into smaller units.

No, the reasoning behind the construction of these giant accelerators is based on the principle of the conservation laws we have been considering in this chapter. In order to see why this is the case, we need to introduce the concept of particle *mass*.

Mass is a **property** of particles. In the same way that a particle might have a certain amount of electric charge, a particle might have a certain amount of mass. A particle's mass determines its resistance to being accelerated, and it also determines its gravitational attraction.

A particle which has mass is called a *massive* particle. We use this term "massive" in everyday language, but we are often rather careless in using the word to describe the size of objects, for example: "The cruise liner is massive". However, physicists use the term "massive" in the very strict sense of describing an object "which possesses mass" (there is also the related word "massless" for describing particles which do not possess mass).

As an example, a neutron star can be about the size of a city, but it is much more massive (i.e., it possesses far more mass). A teaspoon of neutron star material would weigh a billion tons. So "massive" is not a measure of size. With this in mind, remember that all elementary particles should be considered as being infinitely small points with no size. So an elementary particle can be "massive", but have no size! Just think of mass as simply a number – a property – which we assign to a particle.

So a massive particle has a certain amount of mass. And, because of Einstein's famous formula for mass-energy equivalence, $E=mc^2$, we can therefore think of a particle as being composed of a certain amount of energy. With this in mind, the purpose of modern particle accelerators is not to split particles, the aim is to concentrate enough energy into a small volume of space so that a massive particle might be spontaneously created. The more massive the particle, the more energy is required to create it.

As an example, an electron is a massive particle which has a mass of 9×10^{-31} kg. This is such an awkward, extremely small value that the mass is more usually given in terms of its energy equivalent: 0.5 MeV (where "MeV" represents a million electronvolts – remember the electronvolt is a unit of energy which was introduced earlier). So in order to produce an electron from a collision, particles would have to be accelerated to tremendous speeds, thus giving them sufficient kinetic energy. The aim is that the act of compressing enough energy into a small volume of space would allow a massive particle to be spontaneously created from the collision.

So particle accelerators are not "atom smashers" at all – they are particle creators. And the effectiveness of this approach was clearly demonstrated when the Higgs boson was recently created and detected by the LHC with a mass of 126 GeV (1 GeV being equal to a billion electronvolts).

(At the time of writing – December 2015 – the latest news from the LHC is that another particle has possibly been found at 750 GeV, though we will have to wait until summer 2016 for confirmation).

In a BBC documentary entitled *Dancing in the Dark*, Professor David Charlton who works at the LHC explained how new massive particles might be created in the LHC: "When the protons collide, most of the particles which are produced tend to be low-mass particles, such as the familiar protons and neutrons. But sometimes, very rarely, you produce these much more massive particles, and that is what we are looking for. So we might produce Higgs particles, or maybe we might produce even more massive particles which are ones we don't know about – they would be beyond the Standard Model. These are the particles we are really looking for."

David Charlton goes on to explain how these new particles might be detected: "The idea is we are looking for imbalances of momentum of the event that signifies that there are unobserved particles emitted with high-energy, carried out of the detector. What we are seeing is an absence of something, an **imbalance** of something. It is some particles which we can't observe but we can infer that they are there by looking at the rest of the event."

I want to emphasise Prof. Charlton's use of the word "imbalance". We will return to this concept of balance and imbalance later in this book as I believe it is the key to obtaining new insights about the behaviour of Nature. The concept of balance is essential to understanding the universe. Basically, the universe always has to be in balance, and it resists that balance being upset.

But you will recognise how the techniques described by Prof. Charlton for detecting new particles – detecting imbalances in the energy in and out of the interaction – is exactly the same technique described earlier in this chapter which was used to predict the existence of the neutrino, before it was discovered in 1956. And at the core of this approach lie the crucial laws of conservation: energy and momentum, which must be conserved throughout the interaction.

Symmetry

So now we have considered the conservation laws, and seen how they provide a valuable tool for the particle physicist. However, the conservation laws go hand-in-hand with another vitally important principle.

The other great tool has been the discovery of the importance of symmetry. It might come as a surprise to learn that conservation and symmetry appear to be bound together in a very fundamental way. The connection between symmetry and the conservation laws is one of the most profound connections in physics.

We all recognise symmetry when we see it. For example, a human face has left/right *mirror symmetry*. A snowflake has rotational symmetry:

While we all recognise symmetry when we see it, we will need a more formal definition of symmetry for our purposes. And that definition is that if you apply a transformation to an object (for example, rotating the snowflake by sixty degrees) and the transformed object remains identical to the original object then that represents a symmetry. This definition makes it clear that the snowflake possesses rotational symmetry.

Symmetry plays a remarkable role in physics, at a very fundamental level. In this book, we will be exploring how symmetry provides not only the conservation laws, but also the properties of particles, and the forces between those particles. Nobel Prize-winning physicist Phil Anderson has even stated:

It is only slightly overstating the case to say that physics is the study of symmetry.

In order to explain the importance of symmetry, and the connection with conservation, let us introduce the construct known as the *Lagrangian*. The Lagrangian is a hugely important concept in physics, seemingly revealing something profound about the behaviour of Nature. If we are considering classical mechanics (i.e., the study of the motion of large objects – not including quantum mechanics) then we can define the Lagrangian as being equal to the difference between the kinetic energy and the potential energy of an object. In itself, this is not very interesting. However, it becomes interesting when we consider how the value of the Lagrangian changes when an object moves.

We can calculate the Lagrangian of a moving system throughout its motion to get a series of values. If we then sum all those values we get what is known as the *action*. If we consider the action of our moving system over a period of time, we will find that the value of that action will always be the lowest possible value. This very important principle is called the *principle of least action.*

As an example, consider the following image of the trajectory of a ball:

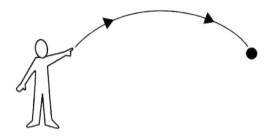

We can calculate the Lagrangian of the motion of the ball (remember: the Lagrangian is the difference between the kinetic energy and the potential energy) at each point, and then we can sum all of those points to calculate the action. The principle of least action tells us that the action will be the smallest possible value for the path taken by the ball. In this case, the smooth path taken by the ball results in the smallest action.

We already know Newton's laws of motion which tell us how an object moves, so why do we need this Lagrangian approach? Well, firstly it can be shown that we can derive Newton's laws of motion from the Lagrangian, so both approaches are equivalent. But the Lagrangian approach seems to be revealing a deeper truth about Nature. Rather than just stating a arbitrary law (e.g., Newton's second law of motion states that the acceleration of an object is proportional to the applied force), the principle of least action seems to reveal a general principle, almost a "desire" by Nature to minimise a certain quantity. As Jerry Marion says in his book *Classical Dynamics of Particles and Systems*:

*In the Newtonian formulation, a certain force on a body is considered to produce a definite motion; that is, a definite **effect** is always associated with a certain **cause**. According to the principle of least action, however, the motion of a body may be considered to result from the attempt of Nature to achieve a certain **purpose**, namely, to minimize the time integral of the difference between the kinetic and potential energies.*

So this underlying purpose appears to provide a rationale as to why objects move the way they do. This generality – this deep truth – means the Lagrangian approach can also be applied to fields, and general relativity, and even to quantum mechanics. As Dwight Neuenschwander says in his book *Emmy Noether's Wonderful Theorem*, the generalizability of the principle of least action to the widest scope of physics, even far beyond mechanics, gives the principle "a depth and versatility not shared by the other mechanical principles."

In order to get useful results from the Lagrangian approach we need to introduce the *Euler-Lagrange equation*:

$$\frac{d}{dt}\left(\frac{\partial L}{\partial \dot{x}}\right) = \frac{\partial L}{\partial x}$$

The equation may look rather daunting but I will attempt to explain it as simply as possible.

The Euler-Lagrange equation is important because it represents the condition when the action is at a minimum value (remember the principle of least action). So the Euler-Lagrange equation can be used to predict the behaviour of Nature (as an example, the Euler-Lagrange equation can be used to derive Newton's laws of motion).

So let us analyze this important equation piece-by-piece.

Firstly, on the right-hand side of the equals sign we find an expression which represents the rate at which the Lagrangian changes when one of the variables describing the motion of the system changes (for example, x might represent the coordinate of a thrown ball as it flies through the air):

$$\frac{d}{dt}\left(\frac{\partial L}{\partial \dot{x}}\right) = \frac{\partial L}{\partial x}$$

The rate of change of the Lagrangian, L, with respect to some variable, x.

We can make this part of the equation a lot simpler if we only consider situations in which the Lagrangian has a symmetry. In a symmetrical situation, as was explained earlier, some value remains unchanged when we apply a transformation. So the rate of change of the Lagrangian in the case of a symmetrical situation would be zero (the Lagrangian does not change when the coordinates change). And so, for symmetrical situations, we can set the right-hand side of the Euler-Lagrange equation to zero. This leaves us with a simpler form of the equation:

$$\frac{d}{dt}\left(\frac{\partial L}{\partial \dot{x}}\right) = 0$$

Now considering the left-hand side of the equals sign, the d/dt part means the *differential* of an expression "with respect to time". This means the "rate of change" of an expression, or – more simply – the speed at which some expression is getting larger or smaller:

The rate of change
with respect to time, t - - - - - $\frac{d}{dt}\left(\frac{\partial L}{\partial \dot{x}} \right) = 0$

Because in this new equation we have set this value equal to zero (on the right-hand side of the equation), this means that the rate of change of the expression in the brackets is zero. Or, put more simply, the expression in the brackets is not changing with time, which means the expression in the brackets is **conserved**.

So, to sum up, this analysis of the Euler-Lagrange equation has shown us that if there is a symmetry in the Lagrangian (i.e., if the value of the Lagrangian is not changed when one of its coordinates changes) **then there will be a conserved quantity**.

As an example, the laws of Nature do not change when we move an object in space. Basically, it does not matter where in space you perform your experiment – you will always get the same result. This is called *space translation invariance* and it was considered in my earlier books. Space translation invariance actually represents a symmetry: remember, a symmetry occurs when we apply a transformation and nothing changes.

As we have seen, the Euler-Lagrange equation indicates that if the Lagrangian does not change with respect to one of the variables of the system then we should be able to find some quantity which is conserved over time. In the particular example of symmetry in space, the conserved quantity (in the brackets) according to the Euler-Lagrange equation is momentum.[7]

So the Euler-Lagrange equation predicts the law of conservation of momentum **because there is a symmetry in space**.

But this is a very mathematical approach. Can we achieve a more intuitive understanding of this principle? Yes, we can.

In his book about the Standard Model of particle physics called *The Theory of Almost Everything*, Robert Oerter presents the example of a skateboarder in a half pipe. The following diagram shows the half pipe oriented in such a way that the skateboarder rolls up and down the pipe, travelling from left to right:

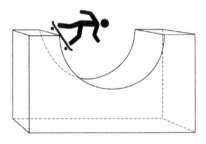

Considering the direction the skateboarder is travelling, there is no symmetry in the half pipe: the pipe drops down, then rises up the other side. Similarly, the skateboarder's momentum is not constant: he is fastest at the bottom of the half pipe, but slows to a stop at the top.

Now let us consider a different orientation of the half pipe. Let us rotate the half pipe. In the following case, the

---

[7] This is because the formula for kinetic energy (used in the formula for the Lagrangian) is $\frac{1}{2}mv^2$, so the partial differential of the Lagrangian with respect to velocity (the partial differential being the conserved quantity) is mv, which is the formula for momentum.

skateboarder again travels from left to right inside the half pipe, but this time the half pipe has symmetry from left to right: the height does not alter:

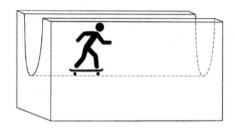

In this case, the momentum of the skateboarder is a constant: he travels at the same constant velocity along the base of the half pipe. In other words, the momentum of the skateboarder is conserved.

So when there is a symmetry in space, momentum is conserved.

The Euler-Lagrange equation can also be used to show that where there is a symmetry in time, energy is conserved. So in our universe – which has both space translation invariance and time translation invariance – we find we have a law of conservation of momentum, and a law of conservation of energy.

However, the full connection between symmetry and conservation was revealed in 1915 by the German mathematician Emmy Noether.

Noether's theorem

Amalie "Emmy" Noether was born in Bavaria, Germany, in 1882. Her father, Max, was a mathematician in the University of Erlangen, and Emmy decided to follow in her father's footsteps by studying mathematics at the university. However, life was not easy for a Jewish female in academia at that time. Indeed, the faculty senate at the university declared that admitting women would "overthrow all academic order". Noether had to work hard throughout her career to overcome prejudice and establish her reputation as a brilliant mathematician.

After earning her Ph.D., Noether published several impressive papers before applying for an academic appointment in the renowned mathematics department at the University of Göttingen. The great mathematician, David Hilbert, who worked at Göttingen, fought for Noether at a faculty meeting: "I do not see that the sex of the candidate is an argument against her admission as an associate professor. After all, we are a university, not a bathhouse."

Hilbert managed to skirt the rules by employing Noether as a guest lecturer. It was shortly after arriving at Göttingen that Noether published the theorem which bears her name.

Noether showed that the previous results from the Euler-Lagrange equation were merely special cases of a more general result. As we saw earlier, the Euler-Lagrange equation indicates that if the Lagrangian is symmetrical with respect to one of its spatial coordinates, then there will be a conserved quantity. Emmy Noether generalised this result by revealing that **if there is any symmetry in the Lagrangian at all, then there is always a conserved quantity**. This result is called *Noether's theorem*.

It is this result of Emmy Noether's which has proven to be of tremendous value for particle physics. If a symmetry is discovered, there will be an associated conserved quantity. The conserved quantity could be something other than momentum or energy. And Noether's theorem works the other way round as well: if a conserved quantity is discovered, there will be an associated symmetry.

So let us end this chapter with a limerick written by David Morin of Harvard University:

As Noether most keenly observed
(And for which much acclaim is deserved),
We can easily see
That for each symmetry,
A quantity must be conserved.

6

GAUGE THEORY

At this point, as we continue or quest for "the forces which hold the atom together", we start to consider the force which holds electrons in their orbit around the atomic nucleus. This is the electromagnetic (or, more simply, the electric) force. In the 18th century it was discovered that positive electric charge attracts negative electric charge, while charges of the same sign repel each other. So, at this point, it would be quite possible to simply state that negatively-charged electrons are held in orbit around the positively-charged nucleus due to the electromagnetic force. And then consider the matter settled, and move on to the next chapter.

However, we can do better than that. Developments in quantum field theory in the second half of the 20th century have provided us with great insights into the fundamental nature of the electromagnetic force. It has been discovered that there is a principle called *gauge theory* which emerges from symmetry and appears to lie at the heart of the behaviour of forces. In this chapter we will examine gauge theory, discovering how it generates forces, and how it predicts the existence of additional fundamental particles.

The electromagnetic force

So electrons are held in their orbit by the electromagnetic force. Electrons have negative electric charge, and are therefore attracted to the protons which have positive electric charge. But why does this attraction exist? Our task in this chapter is to try to get to the fundamentals of this force, analysing how it can reach out over space to attract two objects together.

Crucially, in our attempt to get to the bottom of the electromagnetic force, we have been given a big clue: electric charge is always conserved. This was first discovered by Benjamin Franklin in 1747. If you have a closed system (no charge is allowed to enter or leave the system), then the total amount of charge (positive charge minus negative charge) in the system will not vary with time.

So what can we infer from this conservation of charge? Well, Noether's theorem tells us that whenever we have a conserved quantity, there will always be an associated symmetry. Now we start to see the value of Noether's theorem: we have to find the symmetry inherent in electric charge.

Well, fairly obviously, there is a symmetry between positive charge and negative charge: they are just the two flip sides of the same thing. It appears we have the freedom to instantaneously change every positive charge to a negative charge – and vice versa – without changing the overall situation. This freedom to apply a sudden reversal of a property across the entire universe is called a *global symmetry*.

However, in the mid-1950s, physicists Chen-Ning Yang and Robert Mills realised there was a problem with this universe-wide transformation. They realised that the theory of special relativity prohibited any such global instantaneous transformation across space: no signal can propagate faster than the speed of light. Yang and Mills realised that the change would have to be spread across the universe at the speed of light via a **field**.

Let's imagine we have an electron and we somehow manage to invert its electric charge by making it positive (this is called a *local symmetry* transformation). Essentially, we have taken one corner of the universe and given it a tweak: we have upset the balance of the universe. Remember, the universe only continues to function normally if **every** charge is swapped at the same time. In his book *Deep Down Things*, Bruce Schumm calls this upsetting "a delicate and precise balance" (this theme of balance is becoming a recurring theme). It is as if we have turned-over just one corner of a huge rubber sheet. There will be a ripple of change spreading out across the sheet until the whole sheet turns over. It is just the same for the universe. Our tweak generates a field of change which has to spread out across the whole universe.

Hence, we have discovered that the presence of a local symmetry must result in a field. This necessary connection between symmetry and fields forms the basis of gauge theory which lies at the heart of modern particle physics.[8] As the idea was developed by Yang and Mills, gauge theories are often called *Yang-Mills theories*.

For the case of electric charge symmetry, the gauge field is the electromagnetic field which spreads through space. It is the electromagnetic field which keeps track of charge, responding to any disturbance, and effectively maintaining

---

[8] "Gauge" means a standard of measurement. Gauge theory is based on the principle that it is possible to choose a different standard at each point in space.

the balance of the universe. We will now see how this electromagnetic field is itself formed of fundamental particles.

Bosons

Earlier in the book, we considered the fermions, which are particles with antisymmetric wavefunctions (if you rotate a fermion by 360°, we saw that its wavefunction is multiplied by -1). The elementary fermions were therefore called spin-½ particles. The antisymmetric nature of the fermion wavefunction meant that fermions resisted being together in the same state.

There is another group of particles called *bosons*. In many ways, bosons are the opposite of fermions. If you rotate a boson by 360°, its wavefunction is unchanged (in line with our intuition of how physical rotation should behave). A photon is an example of a boson which has to be rotated once to return to its original state. For this reason, a photon is an example of a *spin-1* particle.

This symmetric behaviour of the bosons means that – contrary to the behaviour of fermions – they like being in the same state. The wavefunction of bosons in the same state does not cancel out (as in the case of fermions), instead it accumulates and becomes larger. This means there is more chance of finding bosons in the same state. Bosons like to congregate together, which explains why many photons like to congregate together to form a ray of light, spreading though space.

All particles can be categorised as either fermions or bosons. It is therefore the most important categorisation of particles.

As was explained in Chapter Three, it is the refusal of fermions to occupy the same state which results in them taking the role of the matter particles, the particles which compose all of the substance in the material world. Cars, trees, water, gases, human beings, are all composed of fermions. The bosons, however, have no resistance to being in the same state, which results in them being intangible and insubstantial. This fact makes it all surprising that – in the *Star Wars* universe – the Jedi Knights decided to construct their swords from light. Lightsabers would be fundamentally incapable of damaging a fermion-based opponent. The moral of this tale is that you should never construct your swords from bosons.

It was James Clerk Maxwell who showed that light was an electromagnetic wave, a disturbance of the electromagnetic field. And it was Einstein who showed that we should think of light as a stream of photons. Hence, we can think of the photon as the boson which forms the electromagnetic field. This makes a lot of sense, as we have just discussed how bosons like to congregate together in vast numbers, which makes them the ideal particle for forming an extended field stretching across space.

So what have we discovered so far in this chapter? Well, in the first section we studied gauge theory, and it was revealed how a local symmetry introduces the necessity for a gauge field which spreads through space. It was stated that, for electric charge symmetry, the resultant gauge field is the electromagnetic field. Now, in this section, it has been explained how the electromagnetic field is itself composed of particles: photons.

In fact, every gauge field is composed of bosons, and these are called *gauge bosons*.

The photon is perfectly suited to the task of being a gauge boson: it is massless and long-range, as suited to the task of keeping track of electric charge across the universe as it is to bringing us light from the distant stars at – well – the speed of light (fairly obviously!).

But there is another interpretation of the role of gauge bosons …

Particles of force

We considered the Feynman diagram of an electron emitting a photon in the chapter on antimatter. We can now consider the diagram of the emitted photon deflecting the path of a second electron:

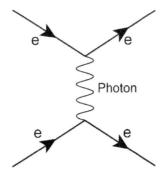

So what does this interaction appear to represent? Well, clearly we could imagine this interaction as representing the electric repulsion between two electrons, two particles with the same electric charge repelling each other. This deflection of electrically-charged particles is the result of the electromagnetic field, which extends through space to affect the paths of electrically-charged particles. This makes a lot of sense, as we have just discussed how the photon is the boson which constitutes the electromagnetic field.

But, from the Feynman diagram, we can also see that we can think of the photon as transferring force. We have two matter particles (the fermions: the electrons), and we have an intangible boson (the photon) transmitting a force across space between the two electrons. Hence, we can think of gauge bosons as the force particles, responsible for transmitting forces between two material objects made of fermions.

To sum up, at the start of this chapter we saw how gauge theory connected the symmetry of electric charge with the necessity for the electromagnetic field. In the next section we saw that the particle which constitutes a field is a boson, and for the electromagnetic field the boson is the photon. In this section, we have seen that we can think of these bosons as being the particles which transmit forces. Hence, the photon is the boson which transmits the electromagnetic force.

But, most importantly, we have discovered a connection between gauge symmetry and forces. In particular, we have seen how gauge symmetry of electric charge results in the electromagnetic force. This is a hugely important result as it is now believed that all of the four fundamental forces arise from gauge symmetry. Gauge theory lies at the heart of the Standard Model of particle physics.

Digging deeper

It is interesting to note that we have not considered the truly fundamental principle underlying the symmetry of electric charge.

As has been discussed earlier in the book, quantum mechanics predicts that every particle also has a wavelike nature: a wavefunction. And every wave – even the waves on the sea – has an associated *phase*. If two waves meet and they are "in phase" then they will interfere constructively and became larger. Conversely, if they are "out of phase" they will interfere destructively and the result will be smaller.

Quantum mechanics tells us that if we have a group of particles – and their wavefunctions – then we can increase or decrease the phase of all the wavefunctions by a constant amount without changing the overall situation. This is because all that matters is the **difference of the phases** of the wavefunctions, **not the absolute values** of the wavefunctions. The difference between phases is important because it can create interference patterns (as in the famous double-slit experiment).

So this is the true source of the symmetry which leads to the electromagnetic field. We have the freedom to increase or decrease the phase of the wavefunctions of particles without changing the overall situation. However, this transformation has to be applied globally in order not to disturb the overall balance of the universe. In order to ensure the global balance is not disturbed, this requires the electromagnetic field. Hence, this is the true source of the gauge field.

If the **sign** of the phase (not just the amount of the phase) of a wavefunction is completely inverted (positive to negative, or vice versa, as if it is reflected in a mirror) then this has the effect of changing the sign of the charge of the particle. So, as was discussed earlier, this is how we have the freedom to completely invert the sign of the charges of a group of particles without changing the overall situation.[9]

And it is this "gauge symmetry" – this freedom to rotate the phase of a particle's wavefunction – which leads to the law of conservation of electric charge (remember: Emily Noether said that each symmetry implies a conserved quantity).[10]

---

[9] This is called *CP-symmetry*, where the "CP" stands for "charge parity". It is known that there are some violations of this symmetry.

[10] Technically, this type of symmetry, being able to rotate the wavefunction of a particle, is give the term U(1) symmetry. The term

We have now reached the conclusion of an important part of this book. In our discussion of "the forces which hold the atom apart" and "the forces which pull the atom together", we have now considered all the forces involved in forming the overall shape of the atom, i.e., the electron shells.

In this chapter, the fundamentals of the electromagnetic force have been described, the force responsible for holding electrons in orbit. We have seen how gauge theory emerges from the conservation of electric charge, and the associated symmetry. It is gauge theory which predicts the existence of the photon, the particle which actually transmits the electromagnetic force.

Hence we have completely examined the forces governing the outer shell of the atom. It is now time to look inside the atom, and examine the nucleus.

"U(1)" means "rotation in one complex dimension". For more information, see the book *Deep Down Things* by Bruce Schumm.

The strong force

In this discussion of electromagnetic attraction and repulsion, a thought might have occurred to you: as protons have positive charge, how can they be tied together so closely in the nucleus without strongly repelling each other? It turns out that a different, stronger force is required to hold the protons together in the nucleus, and this is the *strong nuclear force*.

However, it turns out that the strong nuclear force does not represent a truly fundamental interaction. The true force is even stronger, and operates on a smaller scale, operating **inside** protons and neutrons. The strong nuclear force is then only a residual force, leaking-out of the protons and neutrons, but still strong enough to hold the protons and neutrons together to form atomic nuclei. So to understand the truly fundamental interaction, we have to go inside the protons and neutrons, to uncover the fundamental force known as the *strong force*.

In our discussion so far, we have only dealt with truly elementary particles, particles which are not composed of any smaller particles. However, it has been discovered that both protons and neutrons are each composed of three smaller particles called *quarks*. Quarks are believed to be truly fundamental particles, i.e., they are not composed of anything smaller.

Protons and neutrons are composed of two different types of quarks called *up* quarks and *down* quarks (there are four other types of quarks with hugely greater masses which can only be formed in high-energy particle accelerators, but they all rapidly decay to up and down quarks).

A proton is composed of two up quarks and one down quark. A neutron is composed of one up quark and two down quarks:

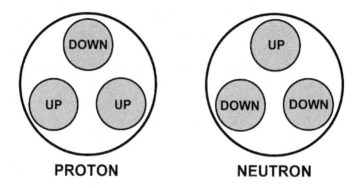

PROTON   **NEUTRON**

But how can this combination of quarks result in the proton having a +1 electric charge, while the neutron has neutral electric charge? For this to be the case for the proton, it appears that twice the charge of the up quark plus the charge of the down quark must equal a +1 electric charge. While for the neutron, it appears that the charge of the up quark plus twice the charge of the down quark must equal zero electric charge. So you might want to remember your high school mathematics and try solving the necessary simultaneous equations to find the answer:

$$2U + D = 1$$
$$U + 2D = 0$$

where U is the electric charge of the up quark, and D is the electric charge of the down quark.

If you do your sums correctly, you will find that – surprisingly – the quarks must have fractional electric charge. The up quark must have a positive two thirds electric charge, while the down quark must have a negative one third electric charge.

However, electric charge is not the charge which holds the quarks together. "Charge" is a generic term which describes how a particle reacts to a particular field (e.g., the electromagnetic field). Only particles with certain charge will react to the field, for example, only particles with electric charge will react to the electromagnetic field. Quarks have a form of charge called *colour charge*. A quark can have either red, green, or blue colour charge, named after the primary colours of light (there is no actual colour involved in any of this – it is just a naming convention).

Just as particles with positive and negative electric charge get attracted together, so quarks with different colour charge get attracted together. But the rule underlying this attraction is quite unusual: quarks are attracted together so that the resultant composite particle has no overall colour. How can this be? Well, consider the mixing of light. A computer (or television) screen is composed of a grid of red, green, and blue dots or stripes, and when all three colours are illuminated, the result is white light (i.e., no colour). This is called additive colour mixing. So quarks are attracted together to form composite particles which are white overall, which can be achieved by attaching a red, green, and blue quark together:

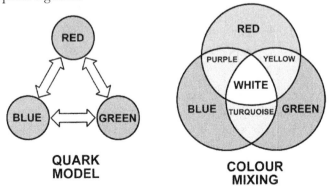

QUARK MODEL

COLOUR MIXING

Hence, this explains how three quarks can be grouped together to forms protons and neutrons.

This principle – that composite particles made of quarks are always colourless – does not just apply to protons and neutrons. For example, we can consider the antiparticle version of a quark (*antiquark*) as having negative colour charge. Hence it is possible for a quark to bind with its antiquark equivalent, for example, a red up quark and a red antidown quark, with the result being colour-neutral. These two-quark particles are called *mesons*. In theory, it is even possible for more than three quarks to form composite particles – as long as the overall result is colourless. In 2015, a *pentaquark* – composed of five quarks – was discovered at the LHC. Again, the pentaquark was an arrangement of quarks which had no overall colour.

One interesting aspect of the behaviour of the strong force is that the quarks become more strongly attracted the further apart they are separated, but at small distances (e.g., inside the proton) there is no longer an attraction, instead the quarks are very loosely bound. This behaviour is called *asymptotic freedom*. As Robert Oerter explains in his book *The Theory of Almost Everything*:

> *As we go to longer distances, the strength of the color force increases. Conversely, at shorter distances, the strength of the color force decreases. It's like putting two fingers inside a rubber band to stretch it. The farther apart you pull your fingers, the larger the force. But if you move your fingers together, the rubber band goes slack.*

We will be returning to consider asymptotic freedom in the next chapter.

Like the electromagnetic force, the strong force emerges as a result of another gauge theory. There is a gauge theory because there is a conserved quantity (colour charge is conserved throughout all interactions), and there is therefore an associated symmetry. Robert Oerter describes this colour symmetry in his book:

> *If we could reach inside every proton and neutron in the universe and instantaneously replace every red quark with a green quark, every green quark with a blue quark, and every blue quark with a red quark, there would be no way to tell that we had done so. The universe would continue on exactly as before the change.*

Note the similarity of this description to the symmetry of charges in the electric force. This is just what you would expect in every gauge theory: if you swap particles in a symmetrical way, you effectively change nothing.

And it is clear we can think of this symmetry transformation as a form of rotation of a quark "colour wheel". When we rotate the colour wheel, red goes to green, green goes to blue, and blue goes to red. This quark colour wheel is shown in the following diagram:

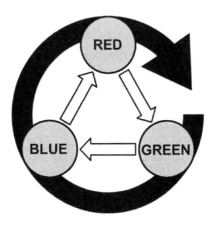

The technical name for this symmetry is SU(3), because there are three different quark colours, three different positions on the colour wheel.[11]

Compare this form of abstract mathematical rotation of particle types on a "symmetry colour wheel" with the actual physical rotation of the snowflake we considered earlier:

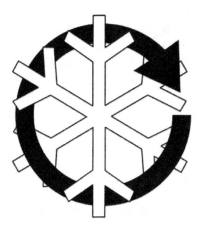

In both cases you can see that it is possible to rotate the object (snowflake, or quark colours) by a certain angle while leaving the situation unchanged. This shows how the principle of symmetry lies at the heart of particle physics.

But what is responsible for this rotation? Well, the strong force is another example of gauge theory, and if we have a gauge theory, we need to have gauge bosons. For the electromagnetic force, the gauge boson was the photon. In

---

[11] The term "SU(3)" means "rotation in three complex dimensions". For more information, see the book *Deep Down Things* by Bruce Schumm.

the case of the more complicated strong force, there is a group of gauge bosons called *gluons*.

The following vertex of a Feynman diagram shows a gluon modifying the path of a quark in much the same way as we have seen a photon modifying the path of an electron:

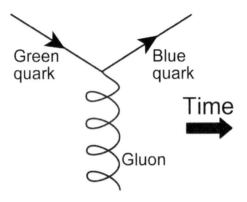

(Note that the symbol for the gluon on a Feynman diagram is different from the symbol for a photon. Yes, they are both gauge bosons, but you can see that the gluon is denoted by a curvy line, whereas the photon was denoted by a wavy line.)

We know that the gluon is the gauge boson for the strong force, so the gluon forms the gauge field which is responsible for keeping track of any changes in local symmetry (any rotation of the local colour wheel) and transmitting those changes across space. We can see in the previous diagram how this is achieved: the gluon has the ability to change the type of a quark. In the diagram, the gluon changes a green quark into a blue quark. So in this way, any local changes of a quark type (any local rotation of the colour wheel) can be effectively transmitted by gluons, changing other quark types accordingly (just like the turning over of a corner of a rubber sheet, described earlier in this chapter).

How many gluons are required? In other words, how many gauge bosons are needed by the strong force? Well, the rule for gauge theories is that you have to count each possible position on the symmetry colour wheel, which in the case of the strong force means we have three different quark colours on the colour wheel. If we call this number n then the number of gauge bosons which are required is given by n^2-1. So, the strong force requires eight gauge bosons, eight different types of gluon.

The theory we considered earlier which describes the electromagnetic force in terms of the interactions between electrons and photons is called *quantum electrodynamics*, or QED. The theory we have just considered which describes the strong force in terms of the interactions between quarks and gluons is called *quantum chromodynamics*, or QCD (where "chromo" means "colour").

The weak force

There are four fundamental forces. So far, we have considered the electromagnetic force, and the strong force. Gravity will be considered in the later chapters of this book. Which means we can now move on to consider the *weak force*.

The weak force is another example of a gauge theory. And if we have a gauge theory, then that means there must be another symmetry. The weak force is based on the symmetry between a down quark and an up quark: a down quark can be changed (rotated on the wheel) into an up quark, and vice versa. There is also a similar symmetry between an electron and a neutrino. When we find two particles connected by a symmetry in this way, they are called a *doublet*.

So the symmetry of the weak force is actually rather simpler than that of the strong force. There are three positions on the "colour wheel" of the strong force, but we see that there are only two positions of symmetry rotation on the symmetry wheel of the weak force:

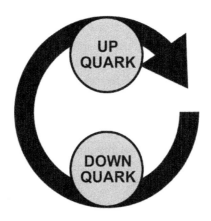

Technically, this is called an SU(2) symmetry (because there are only two positions on the symmetry wheel).

How many gauge bosons are required by the weak force? Well, remember the rule: if n is the number of positions on the symmetry wheel (in this case, $n=2$) then the number of gauge bosons is given by n^2-1. So, the weak force requires three gauge bosons, and these are called the W^+, W^-, and Z bosons. The W^+ has a positive electric charge, the W^- has a negative electric charge, and the Z is neutral (we will soon see why these bosons must have charge).

The following vertex of a Feynman diagram of the weak force shows a down quark changing to an up quark (remember: these two particles form a symmetrical doublet according to the weak force), with the emission of a W^- boson:

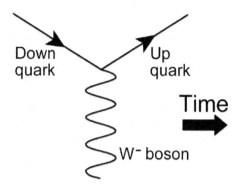

Note the similarity of this diagram representing a weak force interaction with the earlier diagram showing an electron deflected by a photon (according to the electromagnetic force), and also the earlier diagram of a quark deflected by a gluon (according to the strong force). The form of all three diagrams is basically identical, showing that all three gauge theories work on the same principle.

Let us briefly consider those earlier diagrams. According to the electromagnetic force, when the photon "hits" the electron, the photon rotates the symmetry wheel of the electron (actually, the wavefunction of the electron gets rotated – as explained earlier in this chapter). In the diagram of the strong force, when the gluon "hits" the quark, the gluon rotates the symmetry colour wheel of the quark, changing the colour type of the quark. And in the diagram of the weak force, when the W boson "hits" the quark, the boson rotates the symmetry wheel of the quark, changing the down quark into an up quark.

So, for each of the three fundamental forces, we see a gauge boson causing a rotation of a particle's symmetry wheel, usually causing that particle to change into a different particle (the next position on the symmetry wheel).[12]

In all three of the Feynman diagrams, we see a fermion (matter particle: electron or quark) being deflected by a boson. Clearly this, then, is the origin of forces. But we can also see from this discussion that the true root cause of forces is gauge symmetry.

Now let us return to consider the weak force again. The previous diagram showed a down quark changing into an up quark (via the weak force). At this point you might realise we have a problem regarding conservation of electric charge. This is because – if you remember earlier – the down quark has a negative one third electric charge, while the up quark has a positive two thirds electric charge. This means that the difference in electric charge of the quark coming out of the interaction and the quark going into the interaction is one positive unit of electric charge. Essentially, the overall electric charge has increased by one unit – which breaks the law of conservation of electric charge. The only way in which balance can be restored is if the emitted W^- boson possesses one negative unit of electric charge. This explains why the W^- boson has to be negatively-charged (which is why it has a minus sign). Once again, note the importance of the concept of "balance" in this interaction.

Let us now consider the implication of this conversion of a down quark into an up quark via the weak force. Earlier in this chapter, it was described how the neutron is composed of two down quarks and one up quark. Therefore, if one of those down quarks is converted to an up quark, then that

---

[12] Positioning different particle types on a symmetrical rotating wheel is described mathematically by *group theory*. It is possible to transform any element in the group into any other element of the group merely by rotating the wheel. For particle physics, the type of continuous groups which interest us are called *Lie groups* (pronounced "Lee").

neutron is converted into a proton (remember, a proton is composed of two up quarks and one down quark). This conversion of a neutron into a proton is what happens during radioactive beta decay.

The following diagram shows beta decay as a result of the weak force:

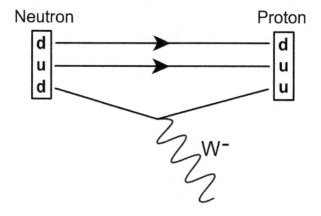

See if you can identify the position of the earlier Feynman diagram for the weak force in this diagram of beta decay. Hence, we can see that forces also provide the mechanism of particle decay.

The emitted W⁻ boson rapidly decays into an electron and antineutrino (as explained in the earlier discussion in Chapter Five of how the neutrino was predicted), and it is the emitted electron which forms the beta radiation.

The Standard Model

With the successful application of gauge theory to particle physics in the 1960s and the 1970s, the field of particle physics was considered to be so mature and well-established that an accurate orthodox model could be constructed. This became known as the Standard Model of particle physics (more usually referred to as simply the *Standard Model*). The Standard Model was developed by many physicists, and has been described by Sheldon Glashow – one of the key architects of the Standard Model – as "a tapestry woven by many hands".

The Standard Model contains a listing of all known elementary particles, together with a deep understanding of their interactions. Experiments have tested the predictions of the Standard Model to great accuracy. The Standard Model is truly an accurate model of the subatomic world.

Because the Standard Model accurately describes the building blocks of our world, it can be used to derive and explain almost all other scientific theories. For example, it is possible to use the Standard Model to derive the laws of electricity and magnetism, thermodynamics, optics, and nuclear energy. It can explain how a star produces energy, and it can explain the functioning of an ant.

So is the Standard Model truly the long-awaited "theory of everything"? Well, not quite. The biggest omission is that the Standard Model says nothing at all about the dominant force in the universe: gravity. It also leaves some other unresolved puzzles, most notably it provides no clue as to the nature of dark matter which supposedly constitutes over 80% of the matter in the universe.

The Standard Model is usually presented as a listing of seventeen elementary particles. We have considered almost all of these particles already in this book:

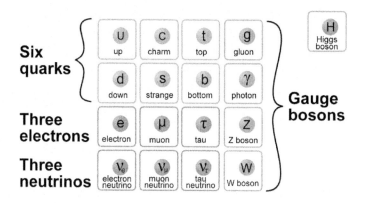

Let us examine the diagram. Firstly, let us consider the twelve fermions in the diagram (the matter particles: the quarks, electrons, and neutrinos on the left of the diagram). We have considered the quarks in this chapter. Electrons have been considered throughout this book (the diagram reveals that there are actually three different types of electron, two of which have vastly greater masses). Neutrinos were considered in Chapter Five (again, the diagram reveals that there are three types of neutrino).

On the right hand side of the diagram we find the five bosons, of which four are gauge bosons. In this chapter we have considered the photon's role in the electromagnetic force. We have also considered the role of gluons in transmitting the strong force (as was explained earlier, there are actually eight gluons in total, but these do not all feature on the diagram), and we have considered the role of the W and Z gauge bosons in transmitting the weak force.

So the only particle we have not yet considered is the Higgs boson …

The unification of forces

On the previous diagram of the particles of the Standard Model, you will see that the Higgs boson stands out on its own. This is because the Higgs boson is not a gauge boson like the other bosons on the diagram – because the Higgs boson does not arise from gauge theory.

Instead, the Higgs boson is the particle associated with the *Higgs field*, and the Higgs field is responsible for giving mass to the elementary particles. Particles which interact more strongly with the Higgs field gain more mass.

We can think of the mass of an object as the object's resistance to being accelerated by a force (or, equivalently, the amount by which the object feels the strength of gravity). In this respect, it is possible to describe the effect of the Higgs field as being similar to the effect of a sea of molasses, slowing the particle down. The Higgs field acts to resist any acceleration of the particle, so the particle gains mass (mass being defined as resistance to acceleration).

What makes the Higgs field interesting from the point of view of the Standard Model is that particle types do not interact with the Higgs field in a symmetrical manner. We can therefore think of the Higgs field as "breaking the symmetry" of particles which would otherwise be identical – in effect, producing the different particles of the Standard Model. As an example, you can see on the previous diagram of the Standard Model that there are three different types of electron: the electron, the muon, and the tau. The only thing which separates these particles is their mass (the tau is heavier than the muon, and the muon is heavier than the electron). If the Higgs field had not given these particles different masses then they would be identical in every way.

Of particular interest is the way the Higgs field appears to have broken the symmetry of four of the gauge bosons: the photon (of the electromagnetic force), and the three gauge bosons of the weak force. It is believed that these four gauge bosons were originally the gauge bosons of a single force, called the *electroweak* force. However, the Higgs field had an asymmetric effect: the end result was to give mass to the weak force bosons, while the photon emerged as a massless particle. As a result, the weak force – with its massive bosons – remained short-ranged and confined to the atomic nucleus, while the massless photons were long-ranged and free to roam the universe. This gave an impression of two separate forces: the long-ranged electromagnetic force, and the short-ranged weak force.

It is believed that this separation of the electroweak force into two separate forces happened in the first moments after the Big Bang, when the universe cooled from extremely high temperature. Perhaps this suggests that forces become unified at high energies?

Indeed, one of the most intriguing outcomes of high-energy particle accelerator experiments is an indication that the effect of the strong force becomes weaker at higher energies, while the strength of the electromagnetic force increases. This seems to suggest that the strengths of the three forces would be the same at extremely high energies. These energies would be a thousand million times greater than can be reached by current particle accelerators, but these conditions would have existed in the first fractions of a second after the Big Bang. Maybe all four fundamental forces – including gravity – were actually one unified force at the time of the Big Bang. After just a fraction of a second after the Big Bang, the universe cooled and the four separate forces would have become "frozen out" to become distinct forces.

7

QUANTUM GRAVITY

Up to this chapter, this book has reflected orthodox physics. The book has been based on the Standard Model of particle physics, which has been experimentally tested to tremendous accuracy time and again. I would imagine all professional physicists would agree with the majority of the book so far.

However, as readers of my previous books will know, my books tend to follow a pattern. The majority of the material presented represents orthodox physics, but I do like to include some original ideas and hypotheses towards the end of the books. This book is no exception. If you wish, you could certainly stop reading this book at this point and feel satisfied that you have read a useful primer on particle physics. But if you are feeling adventurous, and fancy something a bit different (and rather challenging), please press onward. Just bear in mind that much of the material presented in the remaining chapters (especially the material representing my own ideas) should be considered speculative.

However, be reassured that I have confidence in everything which follows.

The Standard Model is undoubtedly a marvellous achievement, but it is not complete: it says nothing about gravity. And that is a fairly monumental omission as gravity is the dominant force which essentially rules the universe.

Einstein's wonderful theory of general relativity is most certainly not the final theory of gravity. General relativity predicts a universe completely spanned by a *gravitational field*. The gravitational field determines how objects move when they are in free-fall, so it effectively defines the force of gravity. However, the gravitational field is a *classical* field – just like Maxwell's electromagnetic field – in that it has no reference to quantum mechanics whatsoever. Einstein's gravitational field is a smooth, continuous surface in all places – not quantized into particles. However, we know that all particles (and fields) have to comply with the principles of quantum mechanics, so there is a need to develop a quantum version of general relativity, with a gravitational field quantized into particles (which, optimistically, have already been given a name: *gravitons*). This would be the final and correct theory of quantum gravity.

Quantum gravity is an active research field in fundamental physics, but definite progress has been painfully slow (non-existent?). The development of a successful theory of quantum gravity is the holy grail of modern fundamental physics.

The biggest challenge for research into quantum gravity is that the two relevant theories – general relativity and quantum mechanics – operate on two completely different scales. General relativity (gravity) is the theory which dominates at large scales, holding stars and galaxies together. Whereas quantum mechanics is the theory which dominates the atomic world. Gravity becomes an irrelevancy for particle interactions – it is simply too weak a force for particles with such small masses. But if we are trying to develop a theory of quantum gravity, that will entail generating equations which describe quantum mechanical behaviour and general relativity **in the same equation**. However, there is simply no way of checking such an equation, or basing such a equation on observations, if we cannot observe anything in the universe for which quantum mechanics **and** general relativity are **both** relevant. The scales of the two phenomena are simply too wide apart: either we observe galaxies and large-scale objects (gravity only – no quantum effects) or we observe particles (quantum mechanics only). It seems there is nothing in-between scales for us to base our equation on.

However, there is just one thing suitable: a black hole.

Black holes are such extreme objects that they can only be completely understood by a theory that must combine both quantum mechanics and general relativity. The event horizon of a black hole is particularly interesting in this respect. Clearly, general relativity is vital to understand the event horizon as general relativity predicts extreme curvature of space at the position of the event horizon (the *Schwarzschild radius*). However, the importance of quantum effects also cannot be ignored at the event horizon. Most notably, Hawking radiation (considered in my second book) is predicted to be generated at the event horizon purely by quantum mechanical effects, and Hawking radiation is vitally important as it is supposed to be the mechanism which eventually results in the evaporation of the black hole.

So in order to fully understand the behaviour of black holes, we will have to develop an equation which describes **both** general relativity **and** quantum mechanics. Black holes really are quite unique objects.

However, even if we are unable to study a black hole closely, our existing knowledge of general relativity and quantum mechanics can provide us with clues about the behaviour of quantum gravity. Starting with quantum mechanics, we know that the uncertainty principle places fundamental limitations of our ability to make accurate measurements of the width of an object. Specifically, if an object has a mass, *m*, then we can only accurately measure the width of that object if its width is no smaller than the object's *Compton wavelength*:

$$l_C = \frac{h}{mc}$$

where *h* is the Planck constant, and *c* is the speed of light. If we try to measure any smaller distance than this then quantum effects – and quantum uncertainty – become dominant.

Remember we are trying to develop a theory of quantum gravity. We have just considered quantum mechanics, so now let us consider general relativity.

General relativity imposes a similar limit on our ability to measure small objects. We know that if we compress a mass to a distance smaller than its Schwarzschild radius then it forms a black hole. So the Schwarzschild radius represents the distance at which general relativity becomes crucial for understanding the behaviour of objects:

$$r_S = \frac{Gm}{c^2}$$

where *G* is the *gravitational constant*.

Remember, we are interested in developing a theory of quantum gravity, so we want to find objects for which both quantum mechanics and general relativity are essential. We therefore need an object with a width equal to its Compton wavelength **and** its Schwarzschild radius. So let us set the formula for Compton wavelength equal to the formula for Schwarzschild radius:

$$\frac{h}{mc} = \frac{Gm}{c^2}$$

Solving for m, we get:

$$m_P = \sqrt{\frac{hc}{G}}$$

This is called the *Planck mass*. Note that the formula for the Planck mass includes the two fundamental constants of general relativity (G and c), and the fundamental constant of quantum mechanics, h.

Now we will take this formula for the Planck mass and substitute it back into the formula for either the Compton wavelength or the Schwarzschild radius (it doesn't matter which one we choose – we have set them both equal in this case). Let's select the Schwarzschild radius:

$$r = \frac{Gm}{c^2} = \frac{G}{c^2} \times \sqrt{\frac{hc}{G}}$$

Square both sides of the equation to get rid of the square root:

$$r^2 = \frac{G^2}{c^4} \times \frac{hc}{G} = \frac{hG}{c^3}$$

Now take the square root of both sides of the equation (to get us back to a formula for distance – not the square of the distance):

$$l_P = \sqrt{\frac{hG}{c^3}}$$

This distance is called the *Planck length*. Again, note that this formula for the Planck length just includes the two fundamental constants of general relativity (G and c), and the fundamental constant of quantum mechanics, h. This is encouraging: it indicates the length is significant for both general relativity and quantum mechanics – just what we were looking for. The Planck length is considered to be the most important distance in quantum gravity research. The suggestion is that the laws of physics at distances smaller than the Planck length could only be described by a joint theory of quantum gravity.

Unfortunately, if you put actual values into the formula you can calculate that the value of the Planck length is an incredibly small distance: 1.6×10^{-35} metres, which is about 10^{-20} times the size of a proton. The fact that this is such a small length is often quoted as the reason why observations of quantum gravity would be so difficult to obtain (particle accelerators would require astronomical energies to probe such small distances).

Rather disappointingly, this represents pretty much the state-of-the-art of what we know for sure about quantum gravity (this, Hawking radiation, and the spin value of the graviton). I told you progress has been slow.

The previous calculation shows that if we want to study the behaviour of quantum gravity, we need to obtain a sample of matter equal to the Planck mass (surprisingly large: about 21 micrograms) and then compress it to its Schwarzschild radius (which, in the case of the Planck mass, is equivalent to the Planck length). Of course, by doing this we are effectively creating a black hole (any mass compressed to its Schwarzschild radius forms a black hole). So what we really need from a theory of quantum gravity is a theory which can describe behaviour at distances smaller than the Schwarzschild radius of a black hole (to be precise, a Planck mass black hole).

At this point, a novel hypothesis which I presented in my earlier books emerges as a potential theory of quantum gravity as it clearly fulfils this requirement. The modified gravity hypothesis (MGH) was first presented in my second book. The whole basis of the MGH is that it predicts novel behaviour inside the Schwarzschild radius. We also saw earlier that any theory of quantum gravity would have to be a theory which can describe the behaviour of black holes – and the MGH fulfils that requirement perfectly.

This is encouraging, but it is only half the battle. To get a full theory of quantum gravity we would need to obtain a quantum version of the MGH, which is what we will consider later in this chapter. But first, please allow me to briefly recap the modified gravity hypothesis.

The modified gravity hypothesis (MGH)

Gravity still remains something of a mystery. We still do not know what happens at the predicted *singularity* at the heart of black holes, or why the expansion of the universe appears to be accelerating. If you read my second book, you will know that it proposed a simple and ingenious modification to the theory of general relativity which solved these – and many more – problems. We sorely need some new insights into how gravity operates at the quantum level, thus providing some pointers to a possible theory of quantum gravity. We will now consider if the modified gravity hypothesis (MGH) proposed in my second book might provide pointers in this direction.

To start, here is a brief reminder of the motivation behind the MGH, and a list of the quite remarkable predictions provided by such a simple idea. I will try to keep this brief, so for the details see my second book.

Firstly, the motivation behind the MGH is that the universe has zero total energy. This statement seems to have become almost accepted as orthodox physics. It is possible for the universe to have zero total energy if gravitational energy is considered to be negative. Richard Feynman considered the relevant equation in one of his 1960s lectures:

If now we compare the total gravitational energy to the total rest energy of the universe, lo and behold, we get the amazing result that the total energy of the universe is zero. Why this should be so is one of the great mysteries – and therefore one of the most important questions in physics. After all, what would be the use of studying physics if the mysteries were not the most important things to investigate?

If we take the implications of Feynman's equation seriously, we find a universe predicted to expand to a certain equilibrium distance, at which point the gravitational energy of the universe will be in perfect balance with the energy contained in the mass of the universe:

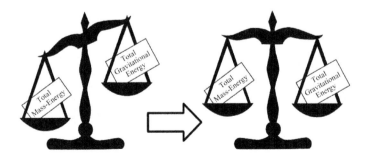

So, once again, we find the importance of the concept of balance in the universe. We have seen throughout this book how balance plays a vital role in constraining and determining the laws of Nature. In our earlier discussion of the conservation laws, we found Professor David Charlton working at the LHC explaining how new particles can be detected from the requirement that the energy going into a particle interaction must be perfectly balanced with the energy coming out. Also, in the previous chapter, it was explained how gauge theory arises from any local disturbance in the balance of the universe, a field being required to correct what Bruce Schumm called "a delicate and precise balance". And in the MGH we again find a universe fighting to restore a balance of energies. As I explained in my second book, it is as though the motivation as to why forces exist at all is because they reconfigure the structure of the universe in order to correct imbalances.

The equilibrium distance, as predicted by the MGH, turns out to be a well-known distance: the Schwarzschild radius. The Schwarzschild radius is best known as the radius of the event horizon of a black hole (if you compress a mass to a size smaller than its Schwarzschild radius, it forms a black hole). But every mass has an associated Schwarzschild radius – not just black holes. What makes the MGH so ingenious is the fact that for almost every object in the universe its Schwarzschild radius is an almost undetectably small distance (the Schwarzschild radius of a human being is about a trillionth of the size of an atom). So, crucially, this results in the MGH predicting an attractive force of gravity which is in agreement with all the well-tested measurements of general relativity.

However, among the remarkable predictions of this simple hypothesis are the following:

- A universe which is spatially flat – without inflation (as a universe which expands to its Schwarzschild radius will be naturally spatially flat).

- A universe with an accelerating expansion – without fine-tuned dark energy.

- A simple solution to the black hole information loss paradox, and a simple explanation as to why the entropy of a black hole is proportional to the area of its event horizon.

It is a very elegant solution to some of the most pressing problems in cosmology. However, as explained earlier, we now want to move on a try to find the implications for quantum gravity.

The asymptotic freedom of gravity

As was explained earlier in this chapter, inside the incredibly small Planck length, it is predicted that the effects of general relativity and quantum mechanics become equally dominant. We really have no idea what reality is like at these scales, though there have been some educated guesses. The most famous guess was by John Wheeler in 1955. Wheeler realised that – at distances smaller than the Planck length – quantum mechanics would play a major role in shaping the spacetime of general relativity. The Heisenberg uncertainty principle would result in random twisting turbulence in the structure of space itself. Wheeler gave this the name *quantum foam* or *spacetime foam*. Unfortunately, in the absence of a theory of quantum gravity, we have no way of testing Wheeler's idea.

The major roadblock in the development of a theory of quantum gravity has been the tendency of infinities to enter candidate theories. The problem in creating a quantum theory of gravity stems from the fact that there is energy contained in the gravitational field connecting the masses. That field energy increases the effective mass of the objects (via $E=mc^2$). But if we increase the mass of the objects, then that increases the strength of the gravitational field, etc. etc. You can imagine how the strength of gravity in this model appears to rapidly shoot off to infinity. And if we have infinities in a physical theory, it shows the theory is flawed. So any potential theory of quantum gravity we might construct would have to possess some mechanism for avoiding these infinities.

The MGH suggests that gravity is not a universally attractive force in all circumstances, but in fact acts to move objects to a certain equilibrium distance (the Schwarzschild radius). Outside the equilibrium distance, gravity acts as a purely attractive force, but inside the equilibrium distance gravity is no longer attractive and no longer acts to bind objects closely together.

Now, put your thinking caps on and try and remember where we have encountered a similar phenomenon earlier in this book.

If you remember back to our discussion of the strong force in the previous chapter, it was explained how quarks become more strongly attracted the further apart they are separated, but at small distances (e.g., inside the proton) there is no longer an attraction, instead the quarks are very loosely bound. This was the phenomenon of asymptotic freedom. A comparison was made to a rubber band being stretched between fingers. **It is clear that asymptotic freedom is a very similar phenomenon to the behaviour of gravity at short distances as predicted by the MGH.** This reveals a possibly intriguing connection between the behaviour of the strong force (QCD) and the behaviour of gravity.

I can quote the Wikipedia article on the development of the theory of asymptotic freedom: "The discovery was instrumental in rehabilitating quantum field theory. Prior to 1973, many theorists suspected that field theory was fundamentally inconsistent because the interactions become infinitely strong at short distances." This sounds uncannily similar to the current puzzling situation regarding the force of gravity, which is predicted to rise to a huge force at small distances, and an infinitely strong singularity in the heart of black holes. The MGH predicts that gravity will no longer be attractive inside the Schwarzschild radius, and I find this similarity to asymptotic freedom intriguing and suggestive. The MGH appears to predict that gravity becomes asymptotically free inside the Schwarzschild radius.

Just as the theory of asymptotic freedom rescued the strong force, maybe the MGH can get the current theory of gravity out of a (very black) hole?

This would maybe suggest that the behaviour of gravitons (the gauge bosons of gravity) has to be similar to the behaviour of gluons (the gauge bosons of the strong force). This is perhaps not such a surprise. The reason given for asymptotic freedom in the strong force is that the gluons can interact with other gluons in complicated ways – giving surprising results. It is known that gravitons must also interact with themselves: gravitons form the gravitational field, the gravitational field contains energy, and energy has a gravitational pull. So "gravity gravitates".

In his book *The Trouble With Physics*, Lee Smolin explains this problem with self-interaction:

> *Gravitational waves interact with each other. They interact with anything that has energy, and they themselves carry energy. This problem does not occur with electromagnetic waves, because though photons*

interact with electric and magnetic charges, they are not themselves charged, so they go right through each other.

Just as self-interacting gluons generate asymptotic freedom, perhaps self-interacting gravitons also exhibit asymptotic freedom?

So the MGH not only explains the universe on the largest scale, but it also points to a new direction for research on the very smallest scale of quantum gravity.

The modified gravity hypothesis (MGH) makes a wide range of predictions which agree with observation.

It explains why gravity is always seen as an attractive force (in agreement with general relativity), but predicts repulsive gravity for the universe as a whole – powering an accelerating expansion. It predicts a spatially-flat universe without inflation. It provides a simple solution to the black hole information loss paradox, and explains the value of black hole entropy. Finally, an asymptotically-free version of gravity arises naturally from the hypothesis, suggesting a new direction for quantum gravity research. It even makes predictions about the behaviour of gravitons.

The MGH is clearly extremely wide-ranging, making predictions below the Planck length, and also at cosmological scales. Hence, the hypothesis starts to resemble a candidate *theory of everything* (or, perhaps more accurately, a *theory of quite a lot*). However, whereas it might require an entire book to describe other theories of everything – such as string theory or loop quantum gravity – the MGH could be described by just a single sentence: "If the universe has zero total energy, then that implies gravity acts to move the masses in a system to the Schwarzschild radius of that system".

We hear a lot about how "string theory predicts gravity". Well, according to the MGH, a zero-energy universe alone predicts Newtonian gravity (for the equations, see my second book). It's all wonderfully simple.[13]

Nobel Prize winner Sheldon Glashow makes another point about string theory: "Superstring theory does not follow as a logical consequence of some appealing set of hypotheses about Nature". However, the MGH most certainly derives from a single logical principle. Indeed, it was built-up from that principle.

While string theory makes few (if any?) firm predictions, the MGH clearly predicts a spatially flat universe. It also predicts a naturally-occurring accelerating expansion of the universe – which is known to be a problem for string theory. The hypothesis also provides a simple solution to the black hole information loss paradox.

I am not suggesting that string theory is wrong. But what I am suggesting is that the modified gravity hypothesis is

---

[13] As a purely speculative thought, it is interesting that the principle that the universe has zero total energy when "viewed from outside" is reminiscent of the colour-neutrality of quark composites produced by the strong force (e.g., protons have zero colour when "viewed from outside"). We can consider mass/energy to be the charge of the force of gravity. Hence, it appears that forces create composite objects which are charge-neutral when viewed from outside: gravity creates objects which are energy-neutral, the strong force creates objects which are colour-neutral, and the electric force creates objects which are electrically-neutral (such as atoms). Is this significant?

more predictive, more testable, and is based on a stronger logical foundation.

The MGH requires no external setting of arbitrary parameters, no landscapes of possible solutions, no multiverses, no hidden dimensions. And unlike other theories – such as string theory, supersymmetry, and loop quantum gravity – the modified gravity hypothesis does not require us to search deeper to discover new structures and particles. Instead, the hypothesis suggests we just need to try harder to make sense of what we already know.

8

THE GREAT UNIVERSE MANIA

To end this book, here is a very peculiar tale.

In the 1930s, a strange fever afflicted the physics department of Cambridge University. But this was no biological virus. This was a strange virus of the mind, a contagious idea so apparently compelling but – at the same time – so cruelly destructive that the reputations of many of the most renowned physicists were permanently damaged.

An article in the journal *Nature* disparagingly called the Cambridge fever "the great universe mania". In this chapter we will examine the cause of this mania, seeing why it caused so much uproar in physics, and trying to understand why so many careers and reputations were damaged. We shall see that among those swept up by the enthusiasm – and destructiveness – of the movement was none other than the great Paul Dirac. It was to be Dirac who made some of the wildest pronouncements during the period of the great universe mania.

Although Dirac received heavy criticism at the time, we shall see that it is possible that Dirac was simply a hundred years ahead of his time.

The numbers of Arthur Eddington

In 1929, the American astronomer Edwin Hubble discovered that the furthest galaxies appeared to be retreating at a faster rate. This appeared to indicate that the universe was expanding, which, in turn, appeared to imply that the universe originated from a small region of space. This was the *Big Bang* theory of the universe, which ushered in a scientific revolution of our image of the universe. Hubble's discovery kickstarted the area of physics research known as *cosmology*, which considers the structure of the entire universe at the largest scales.

The discovery of the Big Bang generated huge ripples of excitement throughout the scientific community. In Cambridge, some of the most notable physicists started to investigate this fledgling field of cosmology. Arthur Eddington was a professor who was one of the first physicists at Cambridge to understand the principles of general relativity. Eddington had created a sensation by showing that light rays bent around the Sun (seen in a photo taken during a solar eclipse), thus proving Einstein's theory of general relativity to be correct. Eddington also became a cosmologist, applying the principles of general relativity to create models of the structure of the universe as a whole.

When Eddington became aware of Hubble's discovery of an expanding universe, he immediately suspected that the *cosmological constant* in Einstein's general relativity equations was the source of the expansion. He then sought to find a mechanism by which the value of the constant might be set to a value which would lead to the observed expansion. But Eddington was also something of a maverick, with an unorthodox style of doing physics. And so the mechanism which Eddington suggested was unlike anything seen before.

Essentially, Eddington realised that we have to use measurement apparatus made of particles (e.g., protons, electrons) in order to measure anything at all. So the size of particles inevitably becomes an intrinsic part of the measurement. According to Eddington, "any apparatus used to measure the world is itself part of the world". So if we want to measure the size of the universe, then the size of particles will inevitably be involved in the measurement result in some way. Eddington took this to mean that there would inevitably always be some fixed relationship between the size of the universe and the size of a particular particle. Hence, although the observed value for the size of the universe might appear to be set to an arbitrary value, it might be possible to find an analytical relationship between the size of the universe and the size of a particle.

If this all sounds a bit crazy, bear with me. It gets crazier.

As an example of the type of peculiar calculation performed by Eddington, we find that the value of the ratio of the strength of the electromagnetic force to the gravitational force is approximately 10^{40}. So what did this value mean? Eddington had no idea. So he decided to see where else this number appeared in physics.

At this point, Eddington entered dangerous territory. He started to use *numerology*. Numerology – as far as physics is concerned – is a largely discredited "science" which essentially involves "working backwards from the answer". The idea is to take some numerical constant and then try to mathematically construct that constant from other known constants, such as pi, or the charge of an electron. If you happen to get lucky – hey – it would appear you have managed to discover some deep secret about Nature. You would have effectively discovered a numerical answer without knowing the underlying theory.

However, this method of "working backwards from the answer" is clearly not as satisfactory as initially constructing a well-founded hypothesis which unambiguously predicts the value of the constant, which is the way good physics is supposed to be done. As John Barrow says in his book *The Constants of Nature*, such a hypothesis would "supply a large and consistent theoretical edifice from which the prediction would follow."

Another problem is that it is possible to play around with numerology without having any expertise in physics, any ability to construct well-founded hypotheses. As a result, numerology has got a bad press as the domain of untalented amateurs who try to promote their dubious (and usually flawed) hypotheses with relentless zeal. Arthur Eddington was also a highly-successful populariser of science, and John Barrow had this to say about Eddington's range of books:

> *The most interesting thing about Eddington's attempts to explain the constants of Nature by algebraic and numerical gymnastics is their enduring effects on the readers of his popular science books. He liked to tell his general readers about his new 'calculations' of the constants of Nature and the overwhelming impression he conveyed was that it might be possible to unlock some of the most deeply hidden secrets of the universe by a little bit of inspired guesswork and numerology. If you noticed that some equations had solutions that lay close to the numbers like 137 and 1840 then you were in business as a rival to Einstein.*
>
> *I believe that Eddington's work, and his widespread popularisation of it in books that sold in huge quantities and continued to be read for more than 60 years after they were first published, inspired a generation of amateurs who dreamed of finding the numerological explanation for the constants of Nature. Every week I receive letters that contain calculations of a sort that owe much to Eddington's style and approach to Nature.*

154

They are characterised by very detailed numerical calculations, a confinement of interest to a small subset of the constants of Nature, and no desire to predict anything new.

If we return to consider Eddington's interest in the huge number 10^{40} (the ratio of the strength of the electromagnetic force to the gravitational force), Eddington realised that this was approximately equal to the square root of the total number of protons in the observable universe, which is believed to be approximately 10^{80}. This became known as the *Eddington number*. However, Eddington was not satisfied by the apparent imprecision in its value. So during a transatlantic boat crossing, Eddington calculated the number to great precision (by hand), eventually proclaiming:

I believe that there are 15,747,724,136,275,002,577,605,653,961,181,555, 468,044,717,914,527,116,709,366,231,425,076,185, 631,031,296 protons in the universe, and the same number of electrons.

Unfortunately, as John Barrow explains in his book, one of the problems with numerology is that "after a while it starts to become addictive". Eddington certainly seems to have been caught by the bug, and it was this contagion which spread around Cambridge University. Eddington started applying his dubious methods to an ever-wider range of constants, often receiving scorn from fellow physicists.

One of Eddington's complicated mathematical formulae supposedly resulted in the value 136, the inverse of the *fine structure constant* (which describes the strength of the electromagnetic force). However, when later measurements placed the value nearer the inverse of 137, Eddington changed his reasoning to suggest that a value of one should be added to his calculation. At this point, some of his critics suggested he should be called "Arthur Adding-one".

After one of Eddington's lectures, two physicists were overheard talking. One said "Do all physicists go off on crazy tangents when they get old?" To which the other physicist replied "Don't worry, you have nothing to worry about. A genius like Eddington may perhaps go nuts, but people like you just get dumber and dumber."

However, I can't help feeling that – as far as physicists are concerned – the main crime committed by numerology is that it has never been successful: it has never been used to correctly predict the value of a constant. Not once. I can't help feeling that if it had ever had any success – however small – then the method would have been rapidly re-evaluated as a perfectly reasonable approach to doing physics. A rather similar re-evaluation seems to have recently happened to multiverse theories which used to be regarded as unscientific but are now being sold as the solution to the current slow progress in physics. If numerology had enjoyed any success then I suspect we would all now be numerologists instead of string theorists. Even a small success can attract a huge amount of attention and research effort.

With this in mind, it might be said that there is nothing particularly unreasonable about an intelligent approach to numerology. As the renowned physicist James Jeans said:

> *Few, if any, of Eddington's colleagues accepted his views in their entirety; indeed, few if any claimed to understand them. But his general train of thought does not seem unreasonable in itself, and it seems likely that some such vast synthesis may in time explain the nature of the world we live in, even though the time may not be yet.*

The return of the amazing Mr. Dirac

In 1937, Paul Dirac was the Lucasian Professor of Mathematics at Cambridge University, one of the most prestigious academic posts in the world. When Dirac made an announcement, the world listened. So it must have come as a great surprise when he announced his decision to change his research area from quantum mechanics to cosmology. As Graham Farmelo says in his biography of Dirac, this involved "refocusing his imagination from scales of a billionth of a centimetre to thousands of light years."

Dirac's first contribution to the topic was written while he was on his honeymoon. It was a 650-word letter to the journal *Nature*. The letter showed that Dirac was not immune to the numerology fervour, as it became clear that Dirac had been largely influenced by the work of Eddington. When Niels Bohr read the letter, he said "Look what happens to people when they get married."

Like Eddington, the letter considered the number 10^{40}, the ratio of the strength of the electromagnetic force to the gravitational force. But Dirac compared this number to the ratio of the radius of the universe to the radius of a proton. The results were very similar, certainly similar enough to convince Dirac that there was a connection. It is certainly unusual to find such a huge number in science, and even more surprising to find approximately the same number arising from two different calculations. As John Barrow said: "There must exist some undiscovered mathematical formula linking the quantities involved. They must be consequences rather than coincidences."

This is called the Dirac *large numbers hypothesis*.

However, it was realised that there was a fundamental problem with Dirac's idea: the universe is expanding. If the

157

universe is expanding, then that suggests that the universe will be larger in the future, and that it was smaller in the past. But Dirac was only using the current radius of the universe in his formula. How could the formula (radius of the universe to the radius of a proton) continue to give the vital 10^{40} number if the radius of the universe was varying?

Dirac provided a fairly astonishing answer. He realised the 10^{40} number could not be maintained in an expanding universe: it was bound to increase. So he suggested that the other magic number – the ratio of the strength of the electromagnetic force to the gravitational force – was also increasing at the same rate. Dirac suggested this increasing ratio would be possible if the strength of the gravitational force was steadily decreasing.

This is surely a sign that Dirac's hypothesis was in deep trouble, with an attempt by Dirac to patch over the problem. If the gravitational force is decreasing, then that suggests the force was much greater in the past. However, it was realised that this increased gravity would have resulted in increased energy output from the Sun, and the Earth would have been much hotter in the past. In 1948, the American physicist Edward Teller showed that in the pre-Cambrian era, 200-300 million years ago, Dirac's workaround would have resulted in the oceans boiling, and life on Earth could not have survived.

So Dirac's hypothesis was falsified, proved to be untrue. Dirac appears to have rapidly lost interest in his idea. But is it now possible to revisit the large numbers hypothesis and show that Dirac was correct in believing that there was a deep underlying connection between the two ratios?

The Dirac connection

The problem with Dirac's approach was that he used the current radius of the universe in order to produce the magical 10^{40} value. In an expanding universe, this value was bound to vary accordingly, thus breaking Dirac's suggested connection.

However, at this point the modified gravity hypothesis (MGH) can step in and provide a solution to Dirac's problem. If you remember, the MGH suggests that the universe has a certain equilibrium radius, and the universe will expand until it reaches that radius. The universe may currently be smaller than the equilibrium radius, or it may currently be larger, but – crucially – **the value of the equilibrium radius does not alter with time**. It is the equilibrium radius which reveals deep truths about the universe – not the current value of the radius, which is an arbitrary value. If we consider the equilibrium radius of the universe then all of a sudden Dirac's hypothesis makes a lot more sense.

If Dirac had used the equilibrium radius instead of the current radius then he would have got the same numerical result. If you remember, the MGH predicts an equilibrium radius for the universe equal to:

$$R_U = \frac{2GM_U}{c^2}$$

where M_U is the mass of the universe.

The ratio of this radius of the universe to, say, the radius of an atom is then approximately:

$$\frac{GM_U}{c^2} \times \frac{1}{R_{ATOM}} \approx 10^{40}$$

So we get the same magical 10^{40} value, but, crucially, all the values in this expression are constant values which do not depend on the current radius of the universe. **So this value does not alter with time**. And if the value does not alter then there is no longer any need for Dirac's horrible workarounds such as a varying force of gravity, which effectively falsified his hypothesis.

It is important to note that this is no longer numerology as practised by Eddington and Dirac. This is a prediction of a well-founded (though admittedly highly-speculative) hypothesis, constructed from first principles, a hypothesis which makes predictions and agrees with known measurements. Dirac's large numbers hypothesis was based on his conviction that these numerical coincidences were revealing some deep – but unknown – truth about the universe. It is possible that the modified gravity hypothesis is the truth he was seeking.

The universe as an atom

However, of course, this is only half the story. So far, we have calculated the value for the ratio of the size of the universe to the size of an atom, and found the value equal to 10^{40}. Crucially, we have also just shown that this value does not necessarily have to alter with time. For the next step, if you remember, Dirac considered this 10^{40} value and found it was the same as the ratio of the strength of the electromagnetic force to the gravitational force.

Is this a just a crazy coincidence? Or could the radiuses of the universe and an atom be related to the strengths of the electromagnetic and gravitational forces in some way?

Firstly, considering the atom. What forces control the size of the atom? Well, the strong force is short range, confined to the nucleus, and so does not play a role. And gravity is too weak at these scales. So it is the electromagnetic force which controls the overall shape of the atom, holding electrons in orbit around the nucleus.

Secondly, considering the universe, the electromagnetic force tends to cancel in atoms, positive charge equalling negative charge, so large objects become electrically neutral. Hence, the electromagnetic force does not play a role in shaping the universe. It is gravity which is the dominant force in the overall size of the universe, even though it is by far the weakest of the four forces. Gravity dominates because all mass has the same gravitational charge (there is no such thing as negative mass). Hence, for very large objects, the force of gravity steadily accumulates until it becomes the dominant force.

So the strength of the electromagnetic force controls the size of the atom, and the strength of gravity determines the size of the universe.

Remember we are now dealing with objects which are in an equilibrium state, meaning the forces holding them together are precisely equal to the forces pulling them apart. This results in an equilibrium radius for the object, and it is this equilibrium distance which interests us. The MGH suggests the universe has an equilibrium radius in much the same way that an atom has an equilibrium radius.

Gravity dominates the universe and determines its size in much the same way as the electromagnetic force determines the overall size of atoms. It is as though the universe is a scaled-up atom!

So, as wild and crazy and randomly unconnected as it may have appeared at first sight, it appears that there might well be an underlying logic to the Dirac large numbers hypothesis.

At these two extremes of scale – the atom and the universe – we find similar situations. We find objects in stable, equilibrium situations dominated by a single particular force. This results in simplicity at the two extremes of scale.

Neil Turok considered this simplicity in his 2015 talk at the Perimeter Institute called *The Amazing Simplicity of Everything* (http://tinyurl.com/turoklecture):

The astonishing thing about recent discoveries in physics is that they tell us the universe is surprisingly simple and regular, on the tiniest scale and on the hugest scale. It's only complicated in the middle. To first approximation, the universe is absolutely uniform in all directions. The whole universe is as simple as the simplest atom. If you think about a hydrogen atom, how many numbers do you need to describe an atom? An atom is a pretty simple thing: you have a nucleus, you have an electron going around it, you have the force of electrical attraction between the nucleus and the electron. Well, it turns out to describe the universe you need just one number. That number describes the universe – fewer numbers than you need to describe a single atom. So the universe turns out to be the simplest thing we know.

The celestial spheres

When we look up at the night sky, we see a number of different celestial objects. Perhaps we see the Moon, maybe we see planets, we almost certainly see stars. In other words, we see **spheres**. And this spherical model applies over the largest range of scales: the largest star has a volume 100 trillion times larger than the Earth, but they are both spheres.

And this consistency over scales still applies if we look down to the smallest scale, where we find atoms, more spheres. Even quarks in a proton are distributed in a spherically symmetric way (for this reason, Bruce Schumm called a proton a "subnuclear atom"). A sphere is an object in perfect balance: the forces which pull the sphere together are balanced by the forces which push the sphere apart. Nature likes balance. Nature tries to correct imbalance.

However, our best current theory of gravity predicts two of the most important objects of all are in a state of raging imbalance. Those two objects are black holes, and the entire universe itself. General relativity predicts that black holes are so totally imbalanced that they are crushed down to an infinitely-small singularity at their centre. As far as the entire universe is concerned, general relativity predicts another out-of-control imbalance – this time in the opposite direction: an ever-increasing expansion of the universe until it is infinitely large.

The modified gravity hypothesis restores balance. It eliminates the infinities.

It predicts spheres.

The modified gravity hypothesis predicts black holes which have their mass smeared around their Schwarzschild radius (in practice, the mass would be concentrated in the accretion disk which surrounds every black hole). For the entire universe, the hypothesis predicts an equilibrium radius, with the universe expanding to that equilibrium like a soap bubble.

The modified gravity hypothesis reveals a universe which seeks perfect balance, from the scale of atoms and protons, up to the scale of black holes, galaxies, and the universe itself.

A universe in perfect balance.

FURTHER READING

Deep Down Things by Bruce Schumm
Still the best guide to particle physics for the general reader.

The Making of the Atomic Bomb by Richard Rhodes
Surely the best book on particle physics ever written.

The Force of Symmetry by Vincent Icke
This gem of a book appears to have been completely overlooked. An accessible description of the role of symmetry in particle physics.

The Constants of Nature by John D. Barrow
Includes a section on numerology, Arthur Eddington, and Paul Dirac. A good read.

The Strangest Man by Graham Farmelo
The definitive biography of Paul Dirac.

Introduction to Elementary Particles by David Griffiths
The best intermediate particle physics textbook. Technical.

Angels and Demons by Dan Brown
Get the *Special Illustrated Edition* off eBay or Amazon – it's wonderful. Steer clear of the movie.

ACKNOWLEDGEMENTS

Thanks to Kip Thorne for his helpful answer to an email query.

I loved *Interstellar*.

PICTURE CREDITS

All photographs are public domain unless otherwise stated.

Etching of the *Investigator* is by Geoffrey Ingleton.

Photograph of the Dirac plaque is by Stansilav Kozlovskiy and is provided by Wikimedia Commons.

HIDDEN PLAINSIGHT 6

Why three dimensions?

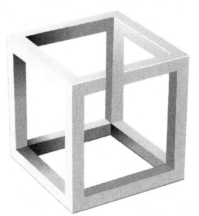

COMING SOON

CPSIA information can be obtained
at www.ICGtesting.com
Printed in the USA
LVOW04s1333190216

475705LV00034B/892/P

9 781519 298874